Plants and

Wildflowers

of

Newfoundland

Plants and Wildflowers

of

Newfoundland

MICHAEL COLLINS

Jesperson Press Limited
1994

Jesperson Press Limited
39 James Lane
St. John's, NF Canada
A1E 3H3

*Cover and Book
Design and Layout:*
Donna Snelgrove

Printing and Binding:
Jesperson Press Limited

The publisher acknowledges a financial contribution from the
Cultural Affairs Division of the *Department of Culture, Recreation and Youth*
of the Government of Newfoundland and Labrador
which has helped make this publication possible.

Appreciation is also expressed to the *Canada Council*
for its assistance in publishing this work.

Printed in Canada.

───────────────────────

Canadian Cataloguing in Publication Data

Collins, M. A. J. (Michael Albert James)

Plants and wildflowers of Newfoundland

Includes index.
ISBN 0-921692-46-3

1. Botany -- Newfoundland.   2. Wild flowers --
Newfoundland. I. Title.

QK203.N5C65 1994   581.9718   C94-950169-7

Cover: *Pitcher Plant*

Michael Collins

TABLE OF CONTENTS

ACKNOWLEDGEMENTS

I have a number of people I would like to thank in the writing of this book. It is appropriate that I should first thank my parents, the late Albert and Mary Collins, for it was their interest in plants, particularly cultivated ones, but also wildflowers, that gave me my initial interest in flowering plants. That initial interest received a significant boost during my school years from my Botany teacher, Rev. Fr. Andrew Alexander, whose enthusiasm for the subject also 'infected' me.

Our botanical training was not just purely theoretical, restricted to the standard lectures and laboratories, for Fr. Alexander believed that it was important for his students to get to know the plants in their natural environments as well. As a result we spent many hours in the field riding our bicycles to sites of local interest. He also encouraged us to conduct biological investigations of the school grounds in our free periods and the laboratory was open at all hours for those wishing to make use of it. His enthusiasm for wildflowers was not just restricted to school hours and I can still remember a trip during one holiday to Selborne, home of that great naturalist Gilbert White, in search of rare flowers the naturalist had found in the area and had documented in his book ,*The Natural History of Selborne*.

When I first came to Newfoundland my interest in things natural was diverted to the winter environment, one that was strange but exciting for someone who came from the relatively mild South of England. I became a winter outdoors enthusiast, as

Michael Collins

is evident from my previous endeavours, with the rest of the year devoted to the pursuit of my passion for tennis. For my more recent interest in wildflowers I have, of all things, to thank our dog! A number of years ago the family debated the addition of a dog to the household, a move I was much against at the time. I was, however, outvoted by the rest of the family. Reluctantly I took pity on the newcomer and started to take him out with me on my winter walks. Walking dogs is, of course, not just a winter activity, so I also took him for walks through the rest of the year. I came face to face with many wildflowers with which, for the most part, I was totally unfamiliar. My earlier enthusiasm for wildflowers was reawakened, so I decided that I might as well start to identify them and purchased a number of (mainland) guides.

I should especially like to thank my good friends and colleagues, Prof. Henry Mann of the Biology Department, Sir Wilfred Grenfell College, and Dr. Peter Scott, Director of the Botanical Gardens at Oxen Pond and Department of Biology, Memorial University, who so freely gave of their time in helping me identify the island's wildflowers and encouraged me throughout the whole project. I owe them a great debt of gratitude, as I do Donna Snelgrove and the members of the staff at Jesperson Press.

INTRODUCTION

Flowers have fascinated human beings since early on in our history on this planet. Even ancient civilizations cultured wild flowers in their communities. For many of these ancient civilizations plants often fulfilled needs other than the purely visual, such as for food, medicines, flavourings, dyes, and even beverages.

Modern human fascination for plants, and especially flowers, is no less than that of former civilizations, and most people now grow up in an environment that includes flowering plants, both in the garden and around the house. But what of the those wild flower-producing plants outside of our homes and gardens? Most of us who know anything about flowering plants probably know more 'domesticated' flowers by name than our own native wildflowers. This is a great shame since we are endowed with a great variety of most attractive wildflowers, many of which also have uses in and around our homes as foods, spices, perfumes, flavourings, and decorations.

The purpose of this book is to enable the resident and tourist alike to identify and learn more about the more common wildflowers of the island. Newfoundland is endowed with a number of rather unique plants, many of which are also included in this book. We have several parasitic plants which obtain their food from other live plants, such as dwarf mistletoe which grows on coniferous trees. We also have several saprophytic plants which lack chlorophyll and break down decaying matter in the soil to

　　　　　　　　　　　　　　　Michael Collins

use as their food. Indian pipe and coralroot are examples of saprophytic plants. Acid bogs which cover some 15% of the island are poor in nutrients and a number of bog plants are specialized to capture and digest insects for the nutrients lacking in these boggy environments. The pitcher plant, the province's floral emblem, is one of the most noticeable of these insectivorous plants, but there are several smaller ones including butterwort, bladderworts and sundews.

Not only do our numerous wildflowers decorate the countryside but many are large enough and long lasting enough to be worthy of a place in our own gardens. An increasing number of people attracted by the beauty of wildflowers in their natural environments have now started to grow them in their own gardens, myself among them. Asters, daisies, goldenrods, fireweed, common ragwort and pearly everlasting are just a few of the native plants which can grace our gardens, and which are certainly not out of place among their horticultural cousins.

Spring is late in coming to Newfoundland, often so late that one could be forgiven for giving up on it altogether! It is not marked by a definite change in the weather or an explosion of green. Often spring is punctuated by late snowfalls, making this season an uncertain transition between winter and summer.

In the typical year, the first spring flowers emerge toward the end of May. I usually look for the white, drooping bells of leatherleaf in the bogs, and the yellow dandelion-like blooms of coltsfoot in waste areas to indicate that spring has finally and definitely arrived at long last! The white flowers of the chuckley pears are usually not far behind those of coltsfoot and leatherleaf.

Spring can be reckoned to last from late May to about the middle of June, with summer following on from then to the end of August. The beginning of August marks the emergence of the last of the late blooming flowers, the goldenrods. The shorter, cooler days of September mark the onset of fall, usually marked by the changing of the pin cherry leaves from green to a pinkish hue. Newfoundland, particularly the east coast and Avalon Peninsula, often has a pleasant fall lacking frosts and early snowfalls, with the consequence that plants can often be found in bloom well into November, and in exceptional years, even into early December. My latest flower record is December 9! Sometimes warm, sunny weather late in fall can fool some of the plants into flowering, that usually only flower in the spring.

Even though we call late-flowering plants fall flowers, all of them have actually begun to flower in mid-summer, often as early as mid-July. Among the hardier of our fall flowering plants are common ragwort, Canada hawkweed, and pearly everlasting, and all can survive moderately heavy frosts and even light snowfalls, so don't give up on the flowers even if there is snow on the ground!

For anyone starting the hobby of flower-watching, fall is to my mind the best time of the year to start since many of the flowers at this time are both large and obvious, and last for such a long period of time. This gives the observer the opportunity to identify and see them time and time again, unlike earlier flowering species which often flower for only a brief period of time.

My interpretation of 'wildflowers' in this book is a very liberal one, unlike similar books which tend to exclude flowering shrubs and trees. Since I see no reason why those interested in identifying or learning more about our native flowering plants should have to consult one book for flowering herbs, and another for flowering shrubs and trees, I have included all common native flowering plants whether they be herbs, shrubs or trees. Having decided to include all common flowering plants in the book, I then had to decide what constitutes a flower.

Many common plants produce flowers which would be recognizable as such only by botanists. Botanically speaking the catkins produced by shrubs and trees, and the heads of grasses, sedges and rushes are flowers, but would not be recognized as such by members of the general public. In deciding what to include in this book I have chosen to include plants which pro-

Michael Collins

duce flowers recognizable as such by a general audience. Such plants also have to be common in distribution across the island, or to be so common locally that their exclusion would be an oversight. I have not included a number of our common native plants that produce flowers which are so diminutive or are hidden from view so that only an ardent naturalist could be expected to notice them. I have also sought the views of a number of knowledgeable botanists on what they would include in, or exclude from, such a book.

The ultimate selection of plants included in this book is my own personal choice. However, I hope to have included the more common flower-producing plants that one would expect to see during the course of the flowering season.

Instructions for use

Different authors have taken quite different approaches to the process of flower identification in their books, and the order in which species are listed in these books. One common approach is to group species according to the colour of the flower, and then to rely on drawings and brief descriptions to identify individual species. One drawback of this colour system is that in a number of different species the colour of the flower can vary considerably, requiring that the same plant be listed in a number of different places in the book. A second drawback of using colour to distinguish between plants is that views differ on what constitutes a particular colour. What may be a shade of pink to one person may be a shade of lilac to another.

Nonetheless, I have chosen to use flower colour as the major basis for identifying plants, but must caution the reader to check sections for related colours if in doubt as to the exact colour of a flower. Check both the yellow and orange sections for flowers with a yellow/orange hue, and the pink and lilac/mauve/purple/violet sections for colours in that range.

Under the heading of Wildflower Descriptions starting on page 18 there is a drawing of each plant as well as a written description of it. Identification keys, starting on page 6, are also provided for each different colour section for specific identification. Some may prefer to use the Key to Wildflowers to make a positive identification, while others may choose to search the drawings in the Wildflower Descriptions section to make the identification. In order to make the latter approach easier the drawings are listed in a definite order, based on leaf characteristics, as listed below:

no leaves (at flowering time)
floating leaves
leaves in a basal rosette
leaves in whorls
opposite leaves
alternate leaves

Since it is impossible to identify plants without the use of some basic botanical terminology a section is included which lists each of the terms used, together with drawings to illustrate some of the more crucial ones. These terms have only been used when they help simplify the process of identification. A list of other books to which the reader can refer for further information has also been included.

Michael Collins

FLOWER PARTS AND LEAF CHARACTERISTICS

Flowers come in all colours, shapes and sizes but all have one thing in common—they are reproductive structures with the function to produce offspring, that is fruits and seeds. Although flowers may seem very different at first glance, they are all derived from the same basic design.

A 'typical' flower, such as say a buttercup or rose, has four concentric rings of parts, namely the pistils, stamens, petals, and sepals. The pistils, the innermost parts, are the female reproductive parts. Each pistil contains an ovary with eggs inside. Arising from the top of the ovary is a slender 'style' which terminates in a 'stigma'. Pollen grains transferred from another flower of the same species are deposited on the stigma, where they produce pollen tubes which then grow down through the style to the ovary. The male gametes then fertilize the eggs inside the ovary to produce seeds.

Surrounding the pistils are the stamens, the male reproductive parts, each of which is composed of a stalk called the 'filament' which is topped by the 'anther' containing the pollen grains.

Outside of the stamens are a number of large, showy petals, whose main function is to attract insects which will pollinate the flower. Sometimes the petals have nectaries at their bases to produce a sugary solution called nectar, which helps to attract insects. Petals may also produce scents to assist in the attraction of insects. In some species the petals may be fused together to produce a bell-like or tubular structure as in blueberries and partridgeberries. Collectively the petals make up the 'corolla'. The outermost ring of parts is the 'calyx' composed of individual sepals. The sepals are usually green and much smaller than the petals. However, in some plants lacking petals the sepals may be large and showy.

Not all flowers have these four sets of parts. Grasses, for instance, are wind pollinated and have no need of large showy petals. As a result their flowers tend to be small and not easily recognizable as flowers to non-botanists. Similarly the catkins on shrubs and trees are also flowers which depend on wind pollination and so also lack showy petals. For the most part the flowers described in this book are insect pollinated and have showy petals or sepals, and are easily recognizable as flowers.

Some plants have unisexual flowers, containing only pistils or stamens. A number of species have the male and female flowers on different plants.

Flowers may be regular, meaning that they can be sectioned in a number of different directions to produce two identical halves (e.g. buttercups and roses) or irregular, when the flower can only be sectioned in one way to produce identical halves (e.g. butter-and-eggs and clovers).

In the composites, members of the daisy family, the most evolutionarily advanced group of flowering plants, what appears to be a flower is in fact an aggregation of many smaller flowers grouped together into a compact flowerhead. In a daisy, for example, the familiar 'flower' with white 'petals' and a central yellow disk, is actually a flowerhead consisting of numbers of two different types of flowers. The 'ray' flowers resemble petals but are sterile, and are arranged in a circle around the disk which is actually made up of many densely packed fertile 'disk' flowers which give rise to the seeds.

There are other examples of flowers which are collections of flowers, rather than individual flowers, such as clovers and crackerberries. These are fully explained in the Wildflower Descriptions.

The flowering plants can be subdivided into two main groups called the *Monocotyledons* and *Dicotyledons*. Monocotyledons generally have long narrow leaves with parallel veins, and the flower parts are in threes or multiples of three. Irises, lilies and daffodils are examples of Monocotyledons. Dicotyledonous plants have broader leaves with network venation, and the flower parts are usually in fours or fives.

Flower Parts

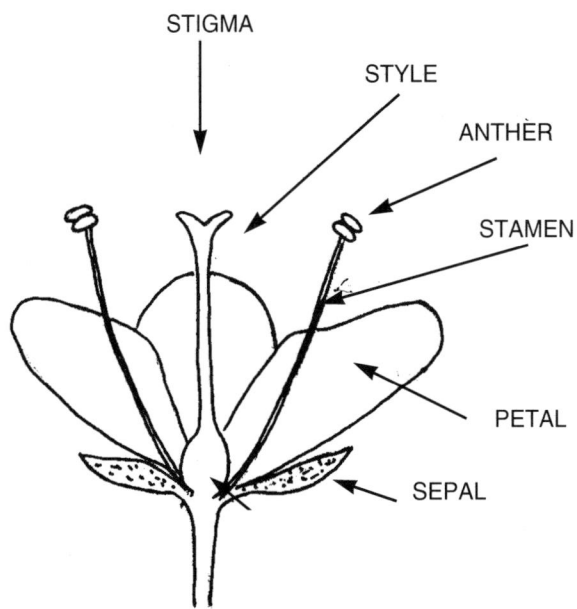

Leaf Arrangements

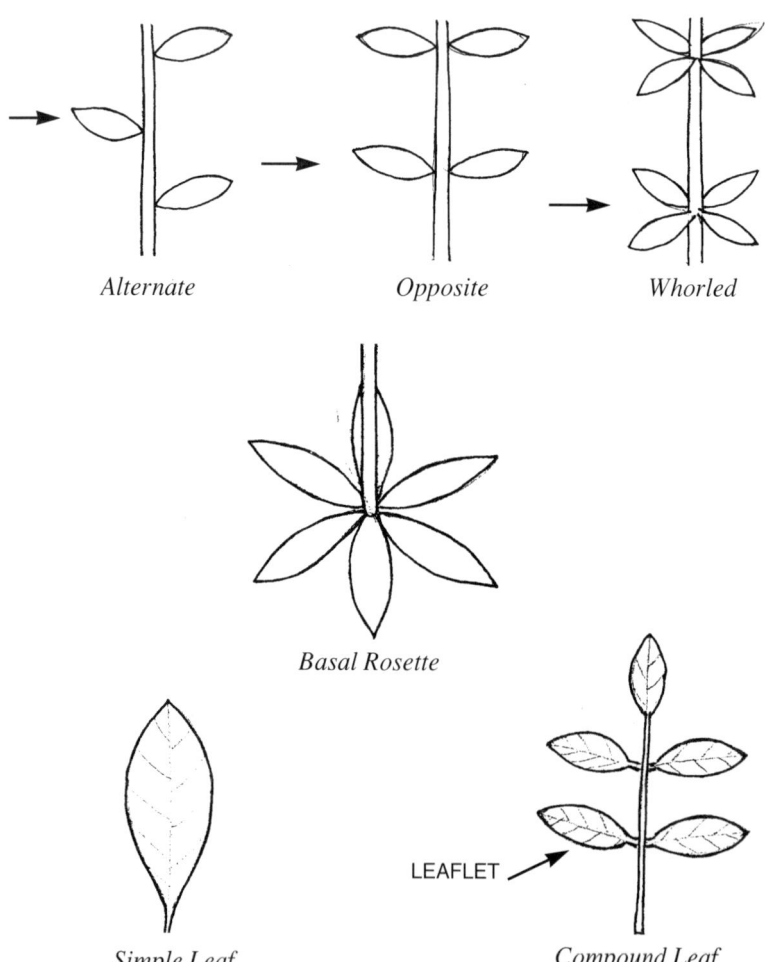

Alternate   Opposite   Whorled

Basal Rosette

LEAFLET

Simple Leaf   Compound Leaf

Michael Collins

Leaf Parts

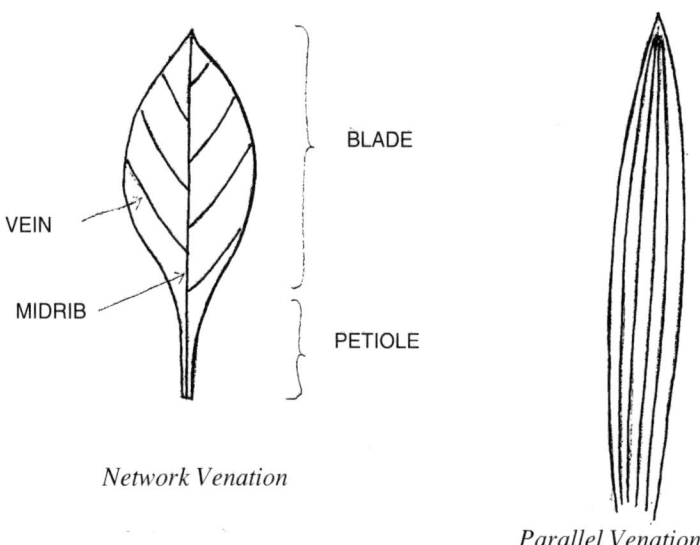

BLADE

VEIN

MIDRIB

PETIOLE

Network Venation

Parallel Venation

Leaf Shapes

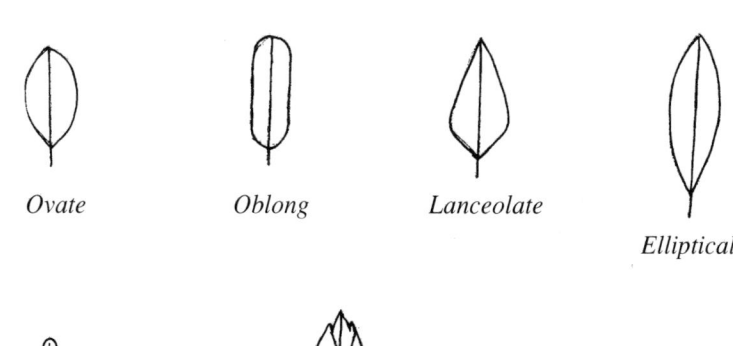

Ovate

Oblong

Lanceolate

Elliptical

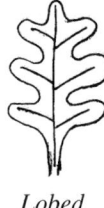

Lobed

Toothed

Plants and Wildflowers of Newfoundland

KEY TO WILDFLOWERS

How to use the Key

Ensure that you are familiar with the criteria we use to identify plants. These are the *leaf and flower parts* and the *leaf arrangements and types,* all shown on pages 3 to 5.

Select the key colour group that corresponds to the colour of the flower you wish to identify:

White — pages 7 to 10
Yellow — pages 10 to 12
Pink and Red — pages 12 to 14
Lilac, Mauve, Purple and Violet — pages 14 to 15
Blue — page 15
Orange — page 15
Dandelion-like plants — page 16
White Flowering Tall Shrubs and Trees — page 17

Under the heading of the correct colour key refer first to item 1 and choose the appropriate option. This will either identify the plant or refer you onwards to another numbered item. If you are referred on again then continue to choose the appropriate options until a positive identification is made.

Once you have made an identification you can see a drawing and full description in the *Wildflower Descriptions* (pages 18 to 140) section of this book and confirm your identification.

There is a glossary (page 141) in case you come across unfamiliar words, and an index (page 146) to help you refer to other parts of the book.

Michael Collins

WHITE FLOWERS

17. Stems covered with bristles, prickles or spines ....Raspberry
                                                                                  and
                                                                   Smooth Blackberry
    Stems not covered with bristles, prickles or spines................................18
18. Seven or more leaflets 19
    Five or fewer leaflets..20
19. Tree with large clusters of flowers; 15 or more leaflets...........Dogberries
    Erect herb of damp places; spike of minute flowers;
    7 to 9 leaflets ....Canadian Burnet
20. Leaves with 5 major subdivisions; leaflets complexly further divided..21
    Leaves with 3 leaflets which are not themselves further divided..........22
21. Terminal cluster of five-petalled flowers ....Bristly Sarsaparilla
    Flowers on separate stalks; four-petalled flowers............Tall Meadowrue
22. Upper leaves simple; lower leaves compound
    and asymmetrical with upper leaflets larger ....Gall-of-the-earth
    Upper and lower leaves compound; leaflets similar in size 23
23. Leaflets not toothed; flowerhead compact
    and globular; irregular flowers..24
    Leaflets toothed; flowers with 5 petals....................................25
24. Flowers and leaves on separate stalks; white
    triangular areas at base of leaflets...White Clover
    Flowers and leaves on same stalks; no white
    triangular areas on leaflets ....Alsike Clover
25. Leaflets resemble maple leaves; leaf base swollen...............Cow Parsnip
    Leaflets strap-like or ovate; leaf bases not swollen...............................26
26. Leaflets strap-like, each with
    3 terminal teeth ....Three-toothed Cinquefoil
    Leaflets ovate with many teeth...27
27. Plant usually has one single flower; height
    usually less than 10 cm...Goldthread
    Plant with many flowers; height usually greater than 10 cm 28
28. Petals much larger than the sepals; flowers
    and leaves on separate stalks...Strawberries
    Petals and sepals roughly equal in size;
    flowers and leaves on same stalks ....Hairy Plumboy
29. Leaves finely divided or with long, narrow lobes 30
    Leaves not divided or lobed 32
30. Leaves fern-like; flowers not scented...31
    Leaves with long narrow lobes; flowers scented ....Musk Mallow
31. Flowers with many white rays and more than
    1 cm across..Scentless Chamomile
    Flowers less than 1 cm across, in flat, compact flowerheads.........Yarrow

Michael Collins

46. Leaves evergreen with orange fuzz beneath ....Labrador Tea
    Leaves deciduous with no orange fuzz beneath 47
47. Leaves clearly toothed; plant of wet areas ....Meadowsweet
    Leaves not toothed; plant of dry areas ....Purple Chokeberry
48. Flowers with ten or fewer rays; flowerheads in
    flat-topped clusters .... Flat-topped White Aster
    Flowers with twenty or more rays; flowerheads
    distributed along upper part of stem ....Panicled Aster

(*See also Striped Toadflax*)

YELLOW FLOWERS

1. No leaves present at flowering time ....Coltsfoot
   Leaves present at flowering time...2
2. Plant of open water with large floating leaves; large
   flowers on stalks arising from below the water surface ....Bullhead Lily
   Plant of land or edges of water; leaves not floating and
   flowers not on stalks arising from below water surface..........................3
3. Leaves all or mostly in basal rosette or arising in
   numbers from the base 4
   Leaves alternate or opposite; usually leaves on upper stem....................9
4. Leaves with prominent teeth or toothlike projections............................5
   Leaves entire with no prominent teeth 6
5. One single flower; flower stalk hollow.....................Common Dandelion
   Several flowers arising from one main
   stalk; flower stalk solid...Fall Dandelion
6. Flowers hanging, not dandelion-like; large
   leaves with parallel venation. ....Clintonia
   Flowers face upward, and dandelion-like;
   leaves do not have parallel venation...7
7. One single dandelion-like flower ....Mouse-ear Hawkweed
   Several dandelion-like flowers arising from one main stalk...................8
8. Leaves only in basal rosette; leaves do not have
   purple mottling ....King Devil
   Several upper stem leaves; basal leaves show
   purple mottling ....Common Hawkweed
9. Leaves opposite 10
   Leaves alternate...12

Michael Collins

10.	Leaves toothed, and over five times longer than wide.........Yellow Rattle
	Leaves not toothed, and length less than five times width....................11
11.	Leaves with transparent 'perforations';
	flowers face upwards; herb ....Common St. John's-wort
	Leaves have no transparent 'perforations';
	flowers hang down; shrub......................................Northern Honeysuckle
12.	Leaves compound with seven or fewer leaflets.....................................13
	Leaves simple, toothed or not, and may be complexly divided.............17
13.	Five or more simple leaflets; woody shrub..................Shrubby Cinquefoil
	Three or fewer toothed leaflets; herb 14
14.	Three equal sized leaflets; stipules at base of leaf................................15
	Leaflets not equal in size; no stipules at leaf base................................17
15.	Leaves toothed; large stipules; petals separate;
	distinct sepals...Rough Cinquefoil
	Leaves not toothed; stipules small; petals not separate..........................16
16.	Flowerheads single, globular and clover-like.........................Hop Clover
	Flowers snapdragon-like and borne in three's..................Birdfoot Trefoil
17.	Three, stalked leaflets; plant spreads by runners ....Creeping Buttercup
	Five to seven stalked leaflets; no runners present ......Common Buttercup
18.	Leaves complexly divided...19
	Leaves simple, with or without teeth or indentations............................20
19.	Leaves finely dissected and fern-like;
	low growing plant ....Pineappleweed
	Leaves finely dissected but not fern-like; very tall plant...Tansy Ragwort
20.	Leaves long, linear and grass-like; more than six times
	longer than broad..21
	Leaves not grass-like; less than five times longer than broad...............22
21.	Has large numbers of upper branches bearing
	numbers of small flowers....................................Lance-leaved Goldenrod
	One main stem bearing yellow and orange
	snapdragon-like flowers ....Butter-and-eggs
22.	Flowers have four flower parts...23
	Flowers have five or more flower parts...24
23.	Flowers large (2 cm or more across) and on
	very long stalks ....Evening-primroses
	Flowers small (less than 1 cm across) and
	short-stalked ....Winter Cress
24.	Flowering parts in five's; plant of wet places 25
	Flowering parts six or more; plant of wet or dry places.........................2
25.	Large flowers; lower leaves kidney-shaped
	and on long stalks...Marsh Marigold

Small flowers; lower leaves roughly
shield-shaped and stalkless ....Spearwort

26. Leaves narrow, elliptical; hundreds of
small indistinct flowers 27
Leaves not narrow or elliptical; flowers usually
distinct and numbering less than twenty in total 29

27. Plant of bogs and wet areas; leaves slightly
clasp stem; flowerheads compact ....Bog Goldenrod
Plant of drier areas; leaves do not clasp stem;
flowerheads on diverging branches...2

28. Leaves without three prominent parallel
veins...Rough-stemmed Goldenrod
Leaves with three prominent parallel veins ....Canada Goldenrod

29. Lower leaves ovate or heart-shaped; blade abruptly
narrowing to very long stalk, and not toothed.....Large-leaved Goldenrod
Leaves not ovate or heart-shaped; stalkless
or short-stalked; toothed or lobed...30

30. Tall plant; leaf has pointed tip; leaves usually slightly
toothed but not complexly lobed;
flowers dandelion-like ....Canada Hawkweed
Small plant; leaf does not have pointed tip; leaves
complexly lobed; flowers not dandelion-like..31

31. Leaf bases extend beyond stem; no clear petal-like parts;
odourless when crushed and not sticky
to the touch...Common Groundsel
Leaf bases do not extend beyond stem; small petal-like rays;
unpleasant odour when crushed and
sticky to the touch...Sticky Groundsel

PINK AND RED FLOWERS

1. Leaves hollow and pitcher-like...Pitcher Plant
Leaves not hollow or pitcher-like..2

2. Single bag-like flower arising from a single stem;
several wide leaves arising from the the base.............Pink Lady's-slipper
Two or more flowers not bag-like, arising from
the main stalk; leaves arising from stem 3

Michael Collins

18.	Plant taller than 1 m; flowers larger than
2 cm and with four petals ....Fireweed
Plant less than 1m tall; flowers smaller than 2 cm 19
19.	Leaves long and narrow without blotches; flowers
bell-shaped and borne at top of stem...................................Bog Rosemary
Leaves not narrow but showing single dark blotches;
small flowers in erect, dense clusters along stem ....Lady's Thumb

LILAC, MAUVE, PURPLE AND
VIOLET FLOWERS

1.	Woody shrub; four-petalled flowers larger than 3 cm;
leaves not fully out at time of flowering......................................Rhodora
Herb or vine; three or four or more flowering parts;
leaves fully expanded at time of flowering 2
2.	Leaves in a basal rosette..3
Leaves alternate or in whorls, but not in a basal rosette..........................4
3.	Usually one single flower; leaves sticky and not toothed.........Butterwort
Several flowers; leaves toothed but not
sticky ....St. Lawrence Bird's-eye Primrose
4.	Leaves in whorls of four or more ....Striped Toadflax
Leaves alternate...5
5.	Compound leaves 6
Simple leaves...7
6.	Three or four leaflets; compact globular flowerheads ....Red Clover
Ten or more leaflets; flowerheads not compact ....Cow Vetch
7.	Leaves, and sometimes stems, bristling with many sharp spines............8
Leaves and stems without sharp spines..9
8.	Spines only on leaves ....Canada Thistle
Spines on both leaves and stem...Bull Thistle
9.	Leaves long, narrow and grass-like...10
Leaves not long, narrow nor grass-like 13
10.	Plant more than 50 cm tall; flowering parts in three's;
flowers more than 3 cm in diameter. ....Blue Flag
Plant less than 50 cm tall; flowering parts
in more than three's, or flowers irregular..11
11.	Flowers with more than ten flowering parts ....Bog Aster
Flowers bell-shaped or snapdragon-like...12

Michael Collins

12. Flowers bell-shaped ....Harebell
    Flowers snapdragon-like ....Striped Toadflax
13. Climbing vine; petals violet with yellow beaks ....Nightshade
    Herb; flowers mauve or violet..14
14. Flowering parts in fives; leaves
    heart-shaped with long stalks ....Marsh Blue Violet
    Flowering parts not in fives; leaves roughly elliptical 15
15. Compact thistle-like flowerheads; leaves have
    tooth-like projections near base. ....Black Knapweed
    Daisy-like flowers with many rays; leaf bases clasp
    stem to a greater or lesser extent 16
16. Very tall plant more than one meter in height; stem
    purplish and bristly; leaf bases clasp stem ....Purple-stemmed Aster
    Plant less than 1 m tall; stem not purplish
    or bristly; leaf bases only slightly clasp stem if at all 17
17. Leaves very long and narrow, often ten times longer
    than broad; floral bracts reflexed ....New York Aster
    Leaves lanceolate, usually five times longer than broad;
    floral bracts not reflexed ....Rough-leaved Aster

BLUE FLOWERS

1. Leaves in basal rosette; leaves and flowerheads
    similar to those of dandelions...Chicory
    Leaves alternate or opposite; leaves roughly ovate;
    flowers with four or five parts..2
2. Leaves opposite; flower parts in fours......................Common Speedwell
    Leaves alternate; flower parts in fives.......................True Forget-me-not

ORANGE FLOWERS

1. Flowers dandelion-like and deep orange
    in colour; leaves in a basal rosette..............................Orange Hawkweed
    Flowers snapdragon-like, yellow and orange or orangey
    in colour; leaves alternate..2

Plants and Wildflowers of Newfoundland   15

2. Flowers bright yellow to orangey in colour and usually
in groups of three; compound leaves ....Birdfoot Trefoil
Flowers mostly yellow in colour, with an orange lower lip;
flowers in circular clusters around the stem;
linear leaves ....Butter-and-eggs

PLANTS WITH
DANDELION-LIKE FLOWERS

1. Flowers blue without stalks ....Chicory
   Flowers yellow or orange 2
2. Flowers orange ....Orange Hawkweed
   Flowers yellow 3
3. Flowers appear before leaves ....Coltsfoot
   Leaves present at same time as flower 4
4. Leaves ascend stem; no basal rosette. ....Canada Hawkweed
   Basal rosette of leaves, with few or no leaves ascending stem 5
5. Basal rosette of leaves with several leaves ascending stem;
   leaves usually with purplish mottling......................Common Hawkweed
   Leaves only in basal rosette with no stem leaves;
   leaves without purplish mottling 6
6. One single flower at the top of the main stalk...7
   Many flowers arise from main stalk...8
7. Flower larger than 2 cm; stalk hollow;
   large toothed leaves ....Common Dandelion
   Flower smaller than 2 cm; stalk solid;
   leaves not toothed...Mouse-ear Hawkweed
8. Leaves with prominent toothlike projections; main stalk
   branched with flowers at the tips of the branches ....Fall Dandelion
   Leaves without toothlike lobes; stalk not branched and
   bearing several flowers at the tip..King Devil

Michael Collins

WHITE-FLOWERING
TALL SHRUBS AND TREES

1. Leaves compound; five or more leaflets 2
    Leaves simple 3
2. Leaves opposite; flowers in terminal spikes ....Red Elderberry
    Leaves alternate; flowers in compact flattish heads ....Dogberries
3. Leaves opposite 4
    Leaves alternate..5
4. Leaves parallel veined; stems often reddish..............Red Osier Dogwood
    Leaves network veined; stems not red.....................Northern Wild Raisin
5. Flowers in dense terminal spikes...6
    Flowers borne singly or in small groups 7
6. Tree; leaves not toothed; stems thick...................................Choke Cherry
    Small shrub; leaves toothed; stems wiry ....Meadowsweet
7. Leaves rounded rectangular often with
    a purplish tinge; petals longer than broad ....Chuckley Pears
    Leaves ovate and without a purplish tinge;
    petals almost round in shape..Pin Cherry

WILDFLOWER DESCRIPTIONS

How to identify a plant

Ensure that you are familiar with the criteria we use to identify plants. These are the *leaf and flower parts,* and the *leaf arrangements and types,* all shown on pages 3 to 5 .

Select the appropriate colour group of plants according to the colour of the flower you wish to identify:

White — pages 19 to 67

Yellow — pages 68 to 100

Pink and Red — pages 101 to 118

Lilac, Mauve, Purple and Violet — pages 119 to 135

Blue — pages 136 to 138

Orange — pages 139 to 140

Under these headings, plants are listed in the following order according to the characteristics of their leaves:

No leaves at flowering time

Floating leaves

Leaves in a basal rosette

Leaves in whorls

Opposite leaves

Alternate leaves

Look through the appropriate category until you can identify the plant. The drawing together with the written description should ensure a positive identification.

There is a glossary (page 141) in case you come across unfamiliar words, and an index (page 146) to help you refer to other parts of the book.

Michael Collins

White Flowers

No leaves or green colour

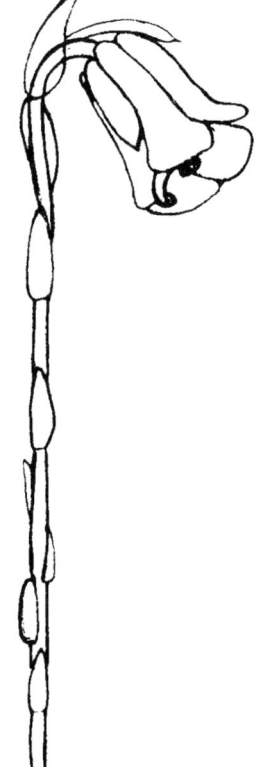

INDIAN PIPE
(Corpse Plant)
Monotropa uniflora
Pyrola family

This is a small, erect white fungus-like plant with no green colour and colourless scale-like leaves that grows to 25 cm tall. Its one hanging bell-like flower gives the whole plant the appearance of an upside down pipe, hence the common name.

It is found in spruce-fir forests during late summer and early fall.

Floating Leaves and Flowers

Flower

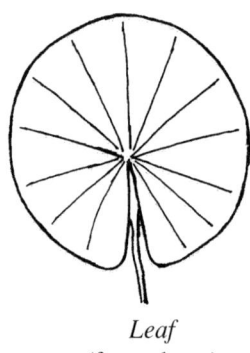

Leaf
(from above)

FRAGRANT WATER LILY
(Sweet Scented Water Lily)
Nymphaea odorata
Water lily family

Leaves and flowers of the Fragrant Water Lily float on the surface of the water, arising from an underground stem rooted in the pond mud. The leaves are rounded with a notch on one side and reddish-green in colour. Its fragrant white flowers are large, 8 cm+ across, with numerous petals. Each flower opens for one day only.

It is found in ponds with muddy bottoms during the late summer and early fall.

Michael Collins

Basal Rosette of Leaves

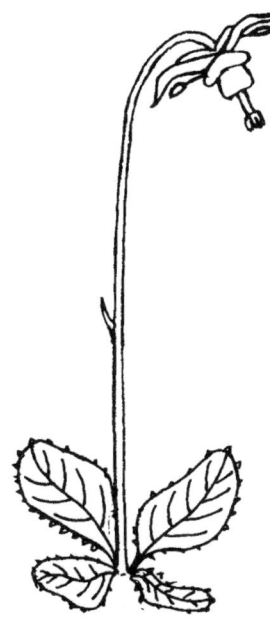

Flower
(from below)

ONE-FLOWERED WINTERGREEN
Moneses uniflora
Wintergreen family

A small, low growing herb of moist woodlands, it has a basal rosette of ovate toothed leaves, and grows up to 10 cm tall. Its one single five-petalled drooping flower has waxy petals, and a prominent green pistil and stigma projecting outward.

It is found in moist woodlands during the late summer.

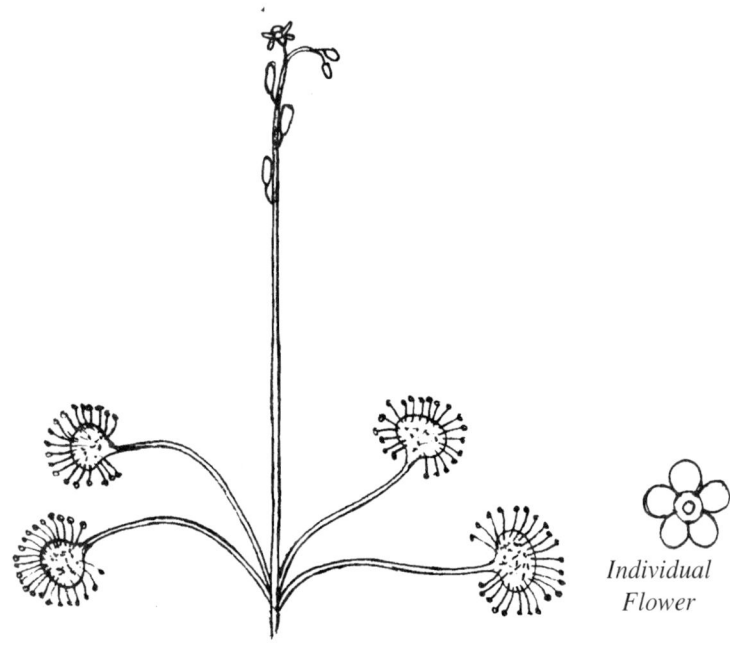

Individual
Flower

ROUND-LEAVED SUNDEW
Drosera rotundifolia
Sundew family

The Round-leaved Sundew is a small low-growing insectivorous plant of bogs, which grows up to 10 cm in height. It has a basal rosette of long-stalked rounded reddish leaves covered with glandular hairs which produce a sticky secretion to trap insects. The leaf curls over to surround the trapped insect which is then digested. Tiny flowers arise at the top of the long stalk and usually only one flower opens at a time.

It is found in wet boglands in the late summer.

(The Spatulate-leaved Sundew (*Drosera intermedia*) has more oval leaf blades.)

Michael Collins

Leaves in Whorls

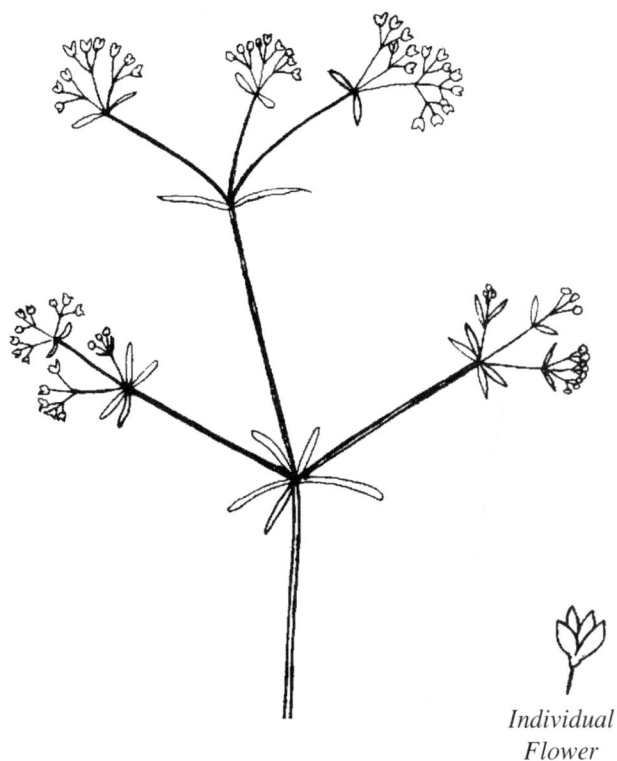

*Individual
Flower*

MARSH BEDSTRAW
Galium palustre
Bedstraw family

Marsh Bedstraw is a low-growing, thin-stemmed straggling herb that leans against other plants for support. The stems are square in cross section and leaves are long and narrow; lower leaves in whorls of four to six, but the upper leaves are opposite. It has terminal clusters of minute four-petalled flowers.

It is found in damp areas during the summer and fall.

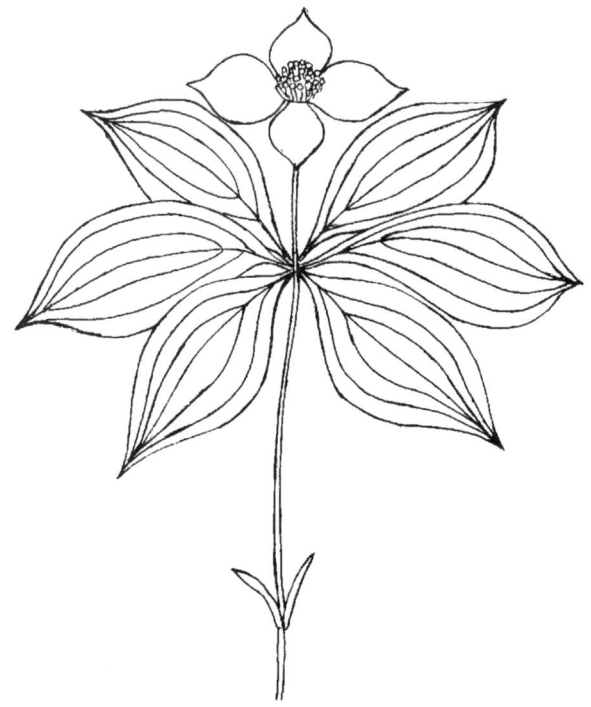

CRACKERBERRY (Bunchberry, Dwarf Cornel)
Cornus canadensis
Dogwood family

This is a low-growing herb of shaded areas, often forming extensive carpets on the forest floor, which grows up to 20 cm tall. Leaves are pointed ovate, parallel-veined in a whorl of six. It has one 'four-petalled' white flower on a short stalk above the leaves, and spreads to 25 mm in diameter. The 'flower' is actually made up of many small, insignificant green flowers in the centre, and the 'petals' are actually modified leaves called 'bracts'. It produces a bunch of edible red berries in late summer.

It is found in forests and woodlands during late spring and early summer.

Michael Collins

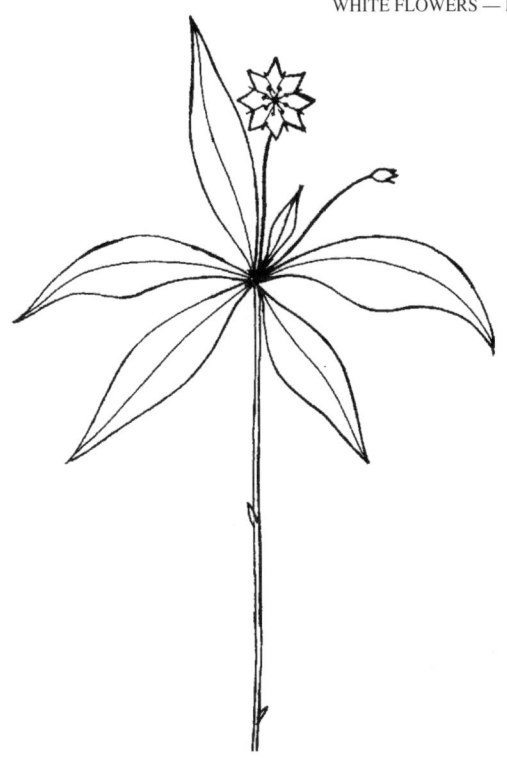

STARFLOWER
Trientalis borealis
Primrose family

This is an erect herb which grows to 20 cm in height. Its flowers and leaves arise from the same stalk with a whorl of, usually, six pointed leaves and bears two 6-8 petalled star-shaped flowers on separate stalks above the leaves.

It is found in forests and woodlands in the late spring.

Opposite Leaves

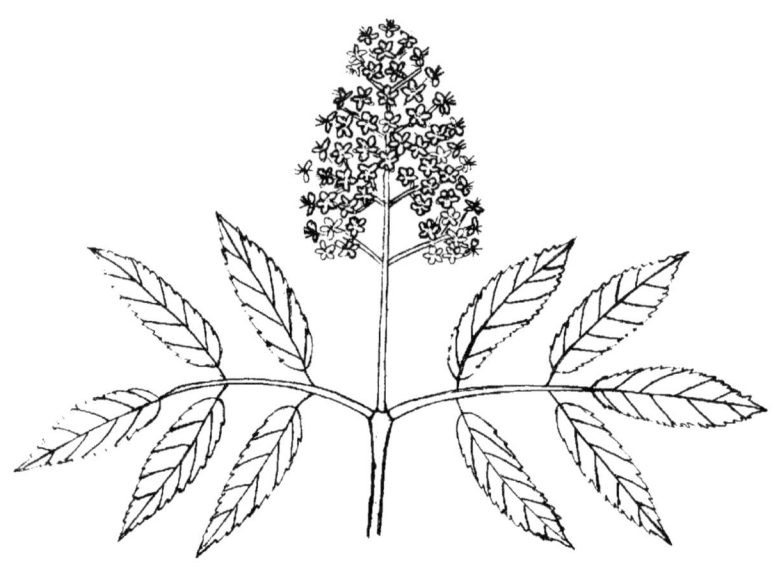

RED ELDERBERRY (Red-berried Elder)
Sambucus pubens
Honeysuckle family

The Red Elderberry varies from a tall shrub to a small tree; branches are brittle. Its leaves are opposite and compound with five to seven toothed leaflets. It has small flowers in erect, terminal clusters which produce clusters of red berries in late summer.

It is found in open forested areas, stream banks and roadsides during the spring and early summer.

(Dogberries have alternate leaves, and Choke Cherry has simple leaves.)

Michael Collins

*Individual
Flower*

RED-OSIER DOGWOOD
Cornus stolonifera
Dogwood family

An erect whip-like shrub, it is usually less than 2 m in height, with few side branches. The older stems are wine-red in colour. Its opposite leaves, are oval-shaped, with parallel veins. If a leaf is gently pulled apart, the two parts are connected by thin gelatinous strands. It has a compact terminal, with a flattened head of four-petalled flowers.

It is found on riverbanks and other wet places during the summer and early fall.

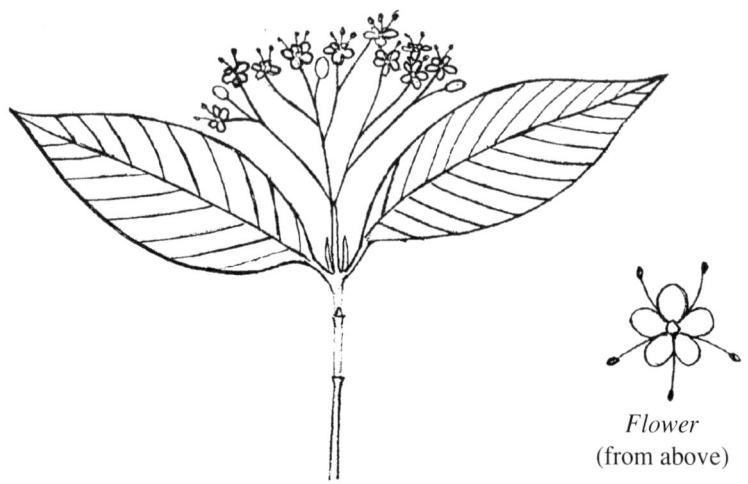

Flower
(from above)

NORTHERN WILD RAISIN
(Witherod)
Viburnum cassinoides
Honeysuckle family

This is a tall shrub, growing up to 3 m tall, with opposite branching and newer branches light brown in colour. Its opposite leaves, are elliptical and stalked with small five-petalled flowers in flattish terminal clusters. Clusters of edible, black berries are produced in the fall.

It is found in open areas and forest clearings during the summer.

Michael Collins

LESSER STITCHWORT
Stellaria graminea
Pink family

The Lesser Stitchwort is a low-growing, thin-stemmed herb which leans against other plants for support, and has leaves opposite and narrow. Flowers are borne at the top of the stem with five petals so deeply cleft as to appear to be ten.

It is found in grassy areas and roadsides during summer and fall.

(Mouse-ear Chickweed has hairy stems and oval leaves.)

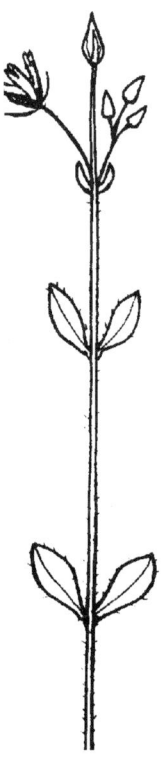

Flower
(from above)

MOUSE-EAR CHICKWEED
Cerastium vulgatum
Pink family

This is a low growing weak-stemmed herb with hairy stems which grows up to 20 cm in height. Its leaves are opposite, oval, hairy and stalkless, and flowers are borne at the top of the stem. The flower has five petals deeply cleft with sepals about as long as the petals.

It is found in waste places during the summer and fall.

(Lesser Stitchwort has narrow leaves.)

Michael Collins

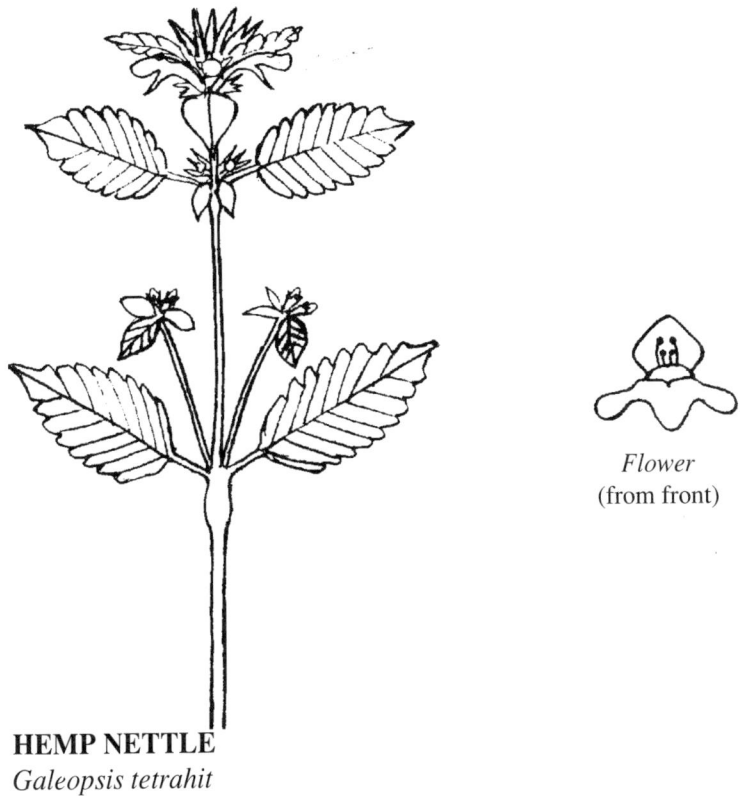

Flower
(from front)

HEMP NETTLE
Galeopsis tetrahit
Mint family

An erect herb, growing up to 1 m in height, the Hemp Nettle has stems that are square in cross section and hairy. Its stems are swollen just below point of leaf attachment. The leaves are opposite, oval, toothed, and hairy, with short stalks. Flowers are in clusters in leaf axils, white or magenta in colour with hood and lips, and the flower envelope has five long bristles.

It is found on roadsides and in waste places during the summer and fall.

(Woundwort doesn't have stalked leaves or swollen stems.)

Plants and Wildflowers of Newfoundland   31

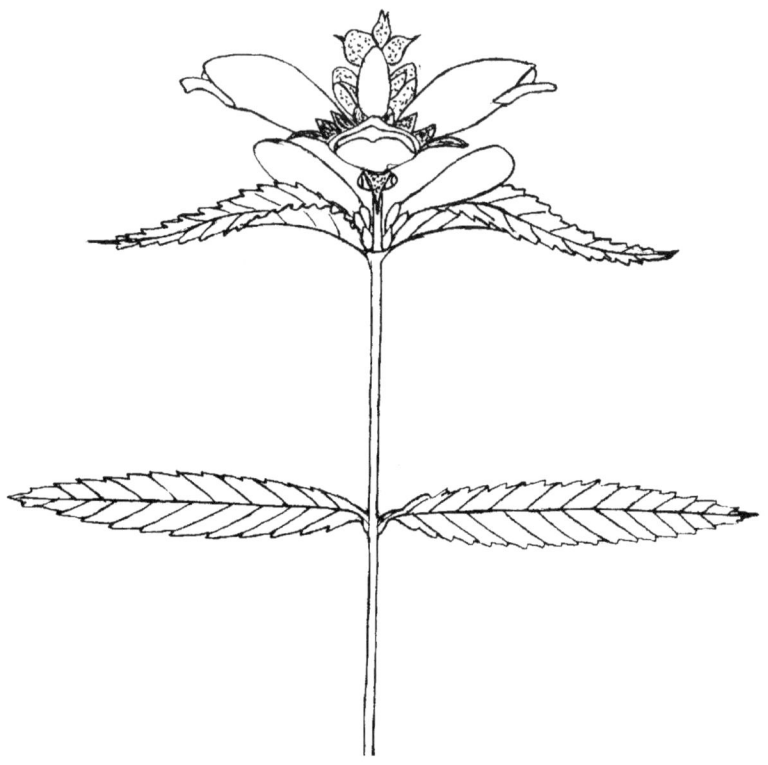

TURTLEHEAD
Chelone glabra
Snapdragon family

    The Turtlehead is a tall herb which grows to 1.5 m with leaves that are opposite, long, narrow and toothed. It has a terminal cluster of large (3 cm) sac-like flowers with upper lip cradling the lower, giving a resemblance to a turtle's head.

    It is found in ditches, riverbanks and other damp areas in the late summer and fall.

32   Michael Collins

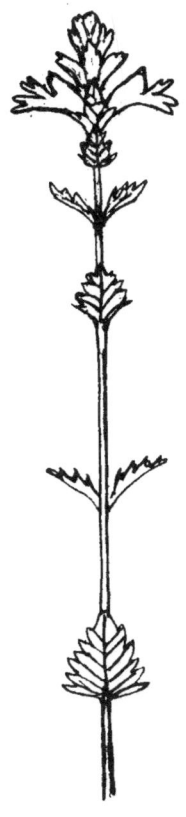

Flower
(from front)

AMERICAN EYEBRIGHT
Euphrasia americana
Snapdragon family

The American Eyebright is a low growing, erect, diminutive herb, less than 25 cm in height. Its stalkless leaves are opposite and tiny with coarse teeth. It has small flowers in leaf axils which are tubular with three lower lips, all forked, and marked with purple lines. It is partially parasitic on grass roots.

It is found on path edges, and grassy areas during the late summer and fall.

Alternate, Compound Leaves

Flower
(from above)

RASPBERRY (Wild Red Raspberry)
Rubus idaeus
Rose family

The Raspberry is a straggling shrub with spiny stems, growing up to 1.5 m tall. Its leaves are alternate and compound, with 3 to 7 toothed leaflets. It has flowers that spread to 20 mm across, but the green sepals are much larger than the white petals. Familiar edible, juicy, red raspberries are produced in the late summer.

It is found in open areas and roadsides.

(Smooth Blackberry has prickles, not spines.)

Michael Collins

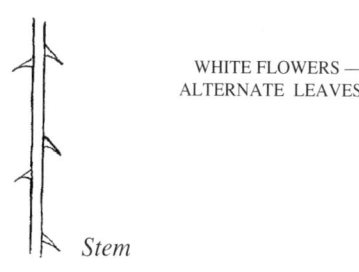

Stem

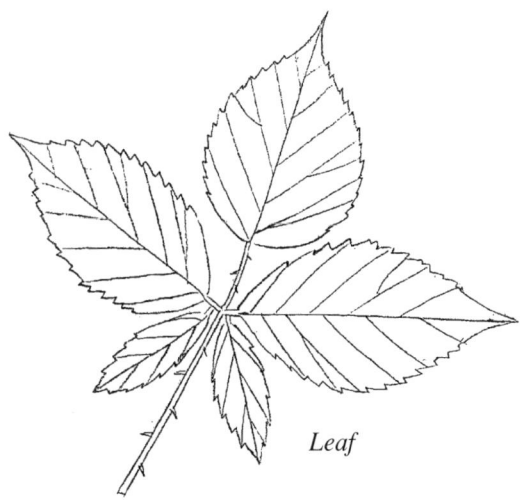

Leaf

SMOOTH BLACKBERRY
Rubus canadensis
Rose family

A straggling shrub, growing up to 1 m tall, with stout prickles on the stems, the leaves of the Smooth Blackberry are alternate, and compound with 3 to 7 toothed leaflets. Its flowers are up to 40 mm across; the white petals are much larger than the sepals. Black versions of raspberry fruits are produced in the fall.

It is found in open areas and on roadsides during the summer.

(Raspberry has spines, not prickles.)

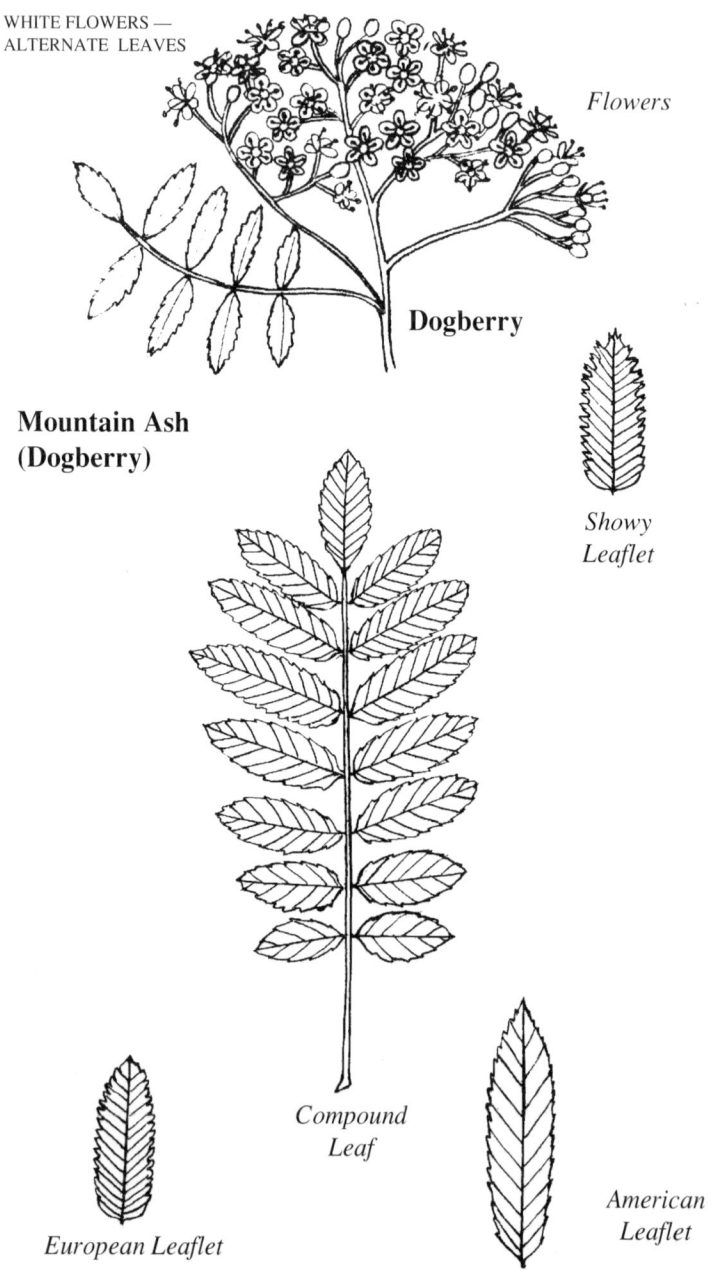

Flowers

Dogberry

**Mountain Ash
(Dogberry)**

*Showy
Leaflet*

*Compound
Leaf*

European Leaflet

*American
Leaflet*

Michael Collins

DOGBERRIES
American Dogberry
Sorbus americana
(American Mountain Ash)
Northern Dogberry
Sorbus decora
(Showy Mountain Ash)
European Mountain Ash
Sorbus aucuparia
Rose family

These are small trees, often with many vertical branches arising near ground level. Their leaves are alternate and compound, with 11 to 17 toothed leaflets. Dogberry trees have five-petalled flowers in dense terminal clusters and produce clusters of orange-red berries in the fall.

They are found in a variety of habitats including forested areas, barrens, roadsides and path edges in the summer.

American: leaflets slightly toothed; each leaflet 4 or 5 times longer than broad.

Northern: leaflets clearly toothed; 2 or 3 times longer than broad.

European: leaflets similar to the Showy but not as markedly toothed. Branches more slender than other species; restricted to some settlements, including the St. John's area.

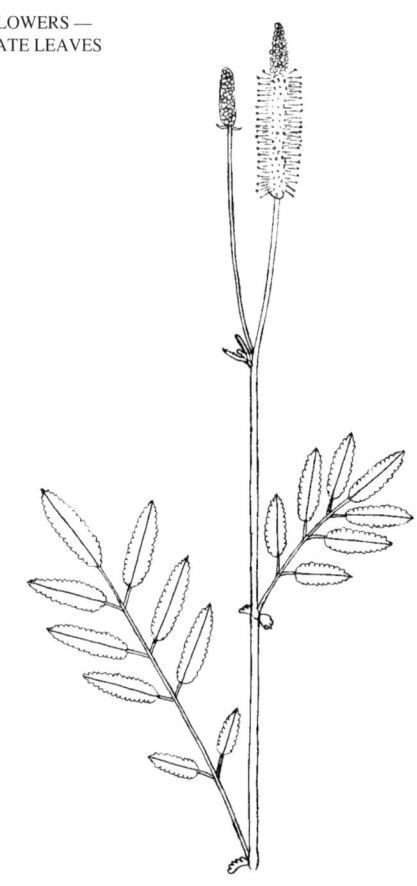

CANADIAN BURNET
Sanguisorba canadensis
Rose family

This is a tall, erect herb, one metre or more in height and has large toothed compound leaves with 7 to 15 leaflets. It has compact terminal spikes of small, white flowers with projecting stamens.

It is found in bogs and ditches in the late summer and fall.

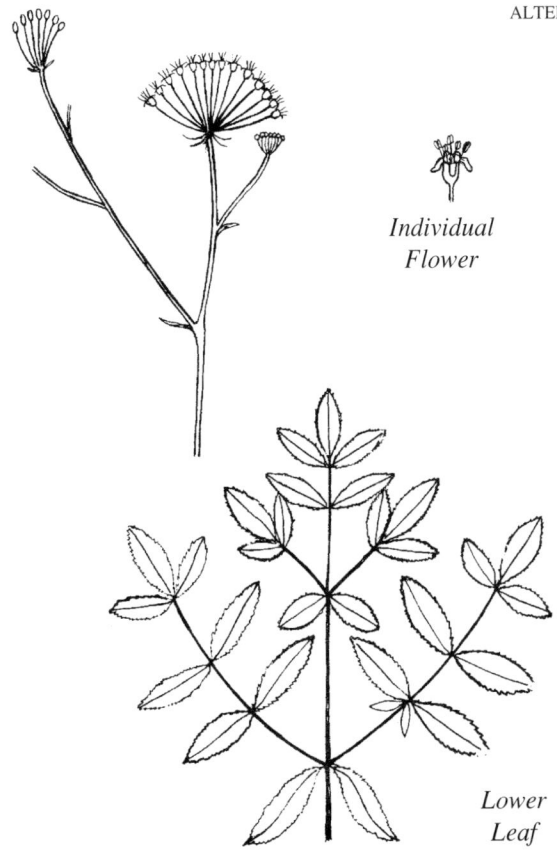

*Individual
Flower*

*Lower
Leaf*

BRISTLY SARSAPARILLA
Aralia hispida
Ginseng family

The Bristly Sarsaparilla, a tall, erect herb which grows to 1 m in height has a very bristly lower stem. The large complex compound leaves have toothed leaflets. Several rounded terminal clusters of small, five-petalled flowers each grow on a long stalk. It produces clusters of small, black berries in fall.

It is found in shady places and woodlands during the summer.

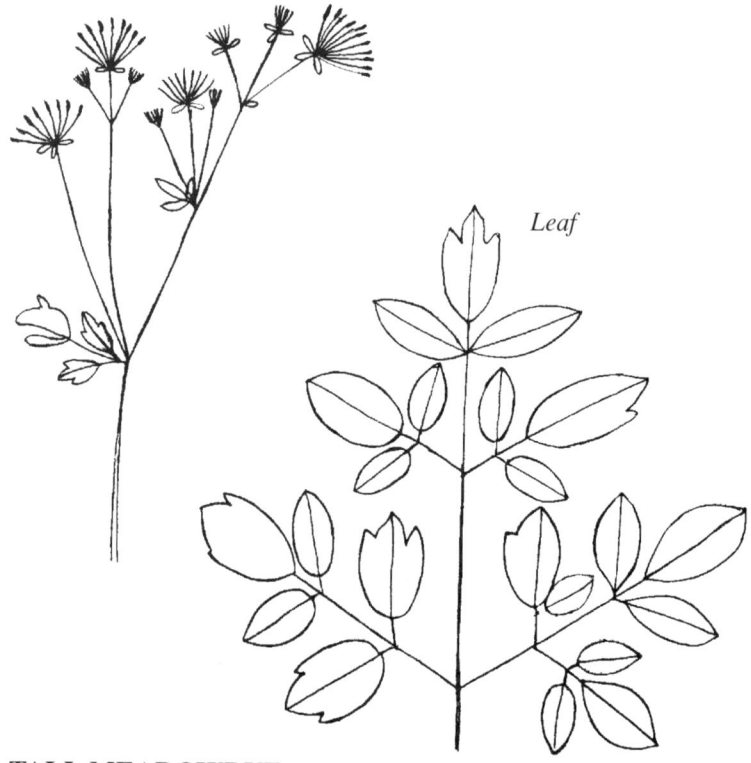

Leaf

TALL MEADOWRUE
(Tall Meadow-Rue, Tall Meadow Rue)
Thalictrum polygamum
Buttercup family

The Tall Meadowrue is a tall herb which grows to 1.5 m in height. Its leaves are alternate and compound, with lower leaves intricately divided; many of the leaflets are three-lobed. The flowers lack petals, but its four small, white sepals are petal-like and the thread-like stamens, topped with yellow, give the appearance of a burst of fireworks.

They are found on stream edges and in other wet areas during the summer.

Michael Collins

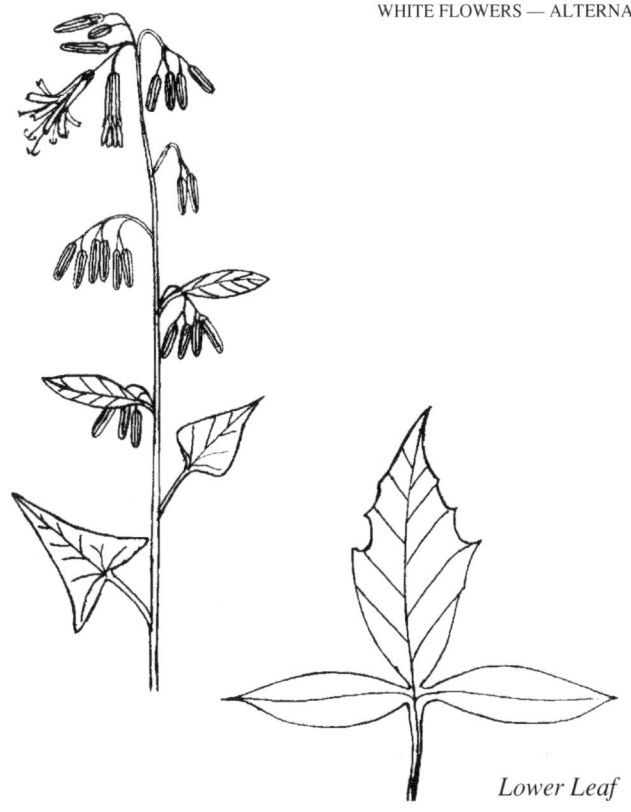

Lower Leaf

GALL-OF-THE-EARTH
(Tall Rattlesnake Root)
Prenanthes trifoliolata
Daisy family

This is an erect herb which grows to 50 cm and has upper leaves which are arrow-headed in shape, while lower leaves are compound with three major subdivisions. Its flowers grow in drooping clusters arising from leaf axils and are creamy white with protective 'sepals'.

It is found in open barren areas in the late summer and fall.

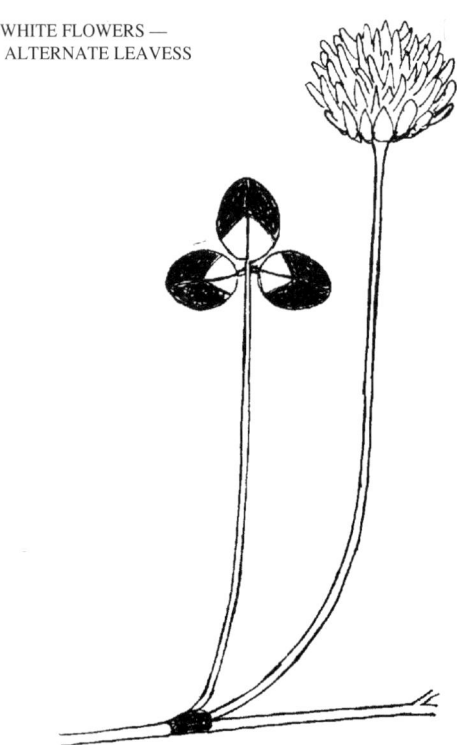

Individual Flower
(side view)

WHITE CLOVER
Trifolium repens
Pea family

This low growing herb is one of our familiar clovers. Its stems are prostrate on the ground with leaves and flowers arising from separate stalks. The compound leaves have long stalks with the familiar three, rounded leaflets (sometimes four and lucky!); leaflets have white triangles at their bases. Flowerhead is spherical and compact, and composed of several dozen individual flowers.

White Clover is found on lawns, grassy areas and path edges during the late spring to fall.

(Alsike Clover has leaves and flowers on the same stalk.)

Michael Collins

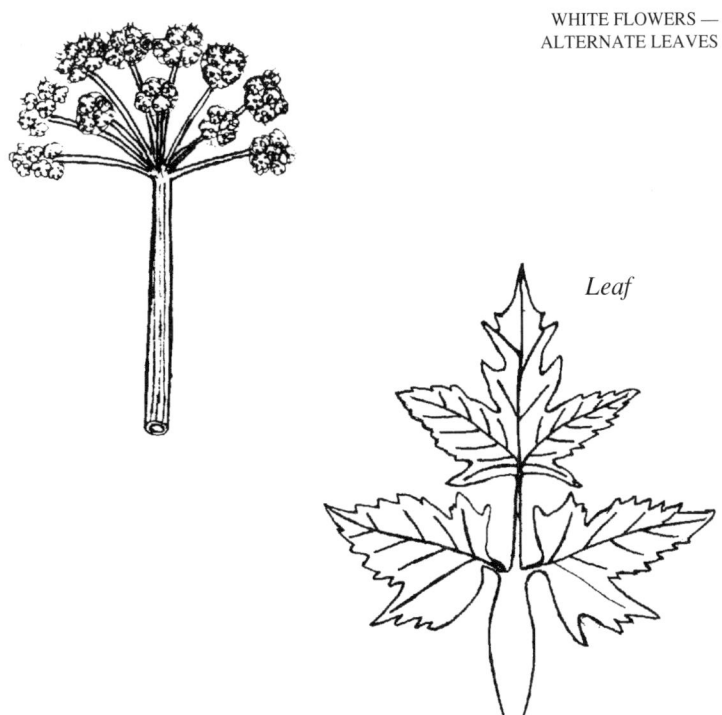

Leaf

COW PARSNIP
Heracleum maximum
Parsley family

Our most common tall, and most massive herbaceous wild-flower often exceeding 2 m in height, has large leaves and is rank smelling. Its stems are hollow and ridged, and the old, dead stems often survive through the next year. The leaves are large and compound, with three leaflets resembling maple leaves; the leaf stalk is inflated where it joins the stem and also partially sheaths it. It has a large terminal flowerhead with many small five-petalled flowers.

The Cow Parsnip is found in ditches, on road edges, favouring areas of good soil, particularly old untended agricultural land during the summer.

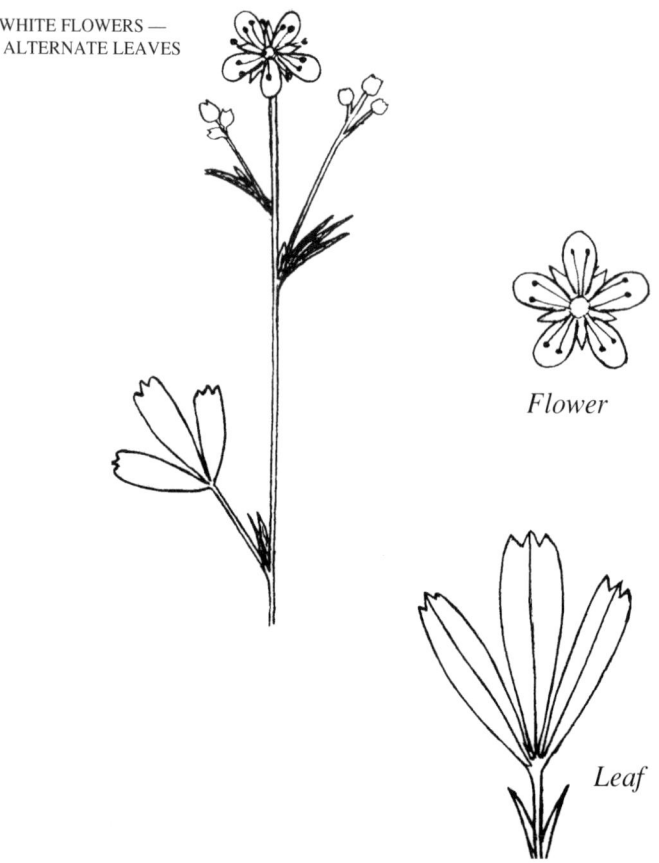

WHITE FLOWERS —
ALTERNATE LEAVES

Flower

Leaf

THREE-TOOTHED CINQUEFOIL
Potentilla tridentata
Rose family

This is a low growing, erect herb which grows to 25 cm in height. Its leaves are compound with 3 strap-like leaflets each with three prominent teeth at the tip, and stipules at the leaf bases. Five-petalled flowers grow in branching clusters.

It is found in open, rocky areas in the late spring to early fall.

(Strawberries have rounded leaflets.)

Michael Collins

GOLDTHREAD
Coptis groenlandica
Buttercup family

The Goldthread is an erect herb with flowers and leaves on separate stalks; the bright yellow 'roots', for which the plant is named, are actually underground rootstocks. It has several evergreen leaves on separate stalks with three, toothed, rounded, leaflets. Each flower is on a separate stalk, while the long narrow 'petals' are actually showy sepals, and the actual petals are small and club-like.

It is found in the spring in shaded forests and woodlands.

WHITE FLOWERS —
ALTERNATE LEAVES

STRAWBERRIES
Fragaria species
Rose family

Strawberries are low growing, spreading herbs; spread by overground runners. It has compound leaves on long stalks with three, rounded and toothed leaflets. Five-petalled white flowers grow in clusters and are up to 2 cm across. The familiar red, juicy edible fruits are produced in summer.

It is found in open grassy areas in the spring and early summer.

Common Strawberry (*Fragaria virginiana*) has the flowers and leaves at the same level, with the fruits rounded.

Wood Strawberry (*Fragaria vesca*) has the flowers above the leaves, and the fruits are conical.

Michael Collins

Flower
(from above)

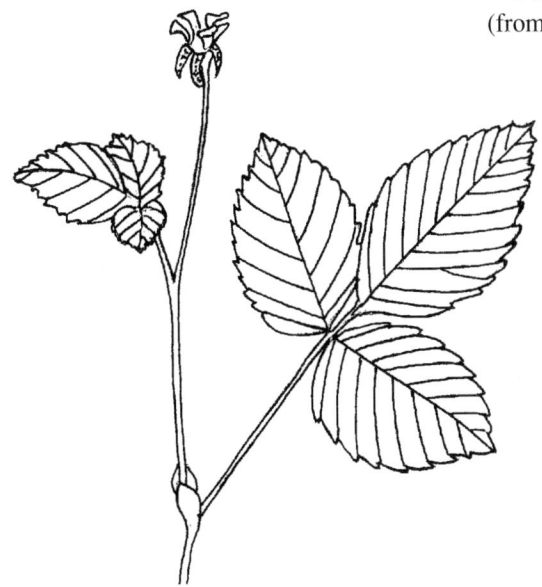

HAIRY PLUMBOY
(Dwarf Raspberry)
Rubus pubescens
Rose family

The leaves and flowers of the Hairy Plumboy grow on vertical branches arising from a trailing stem. It grows up to 50 cm in height. There are no prickles or spines as in other raspberries. Compound leaves have three, toothed, rounded leaflets. It has irregular flowers with five erect petals and five sepals bent downward, and produces dark red raspberries.

It is found in bogs and damp woods in the spring and early summer.

(Blackberries and raspberries have bristles, prickles or spines.)

MUSK MALLOW
(Lords and Ladies)
Malva moschata
Mallow family

The Musk Mallow is a tall erect herb which grows to 1 m. Its leaves are compound and deeply divided into narrow segments. It has large, five-petalled perfumed flowers which spread to 5 cm across with petals notched at tips, and can be either white or pink.

It is found in waste places, fields and roadsides in the summer and early fall.

(Roses are woody with bristles or prickles.)

Michael Collins

SCENTLESS CHAMOMILE
Matricaria maritima
Daisy family

The Scentless Chamomile is an erect, much branched herb, with many daisy-like flowers that have no aroma. It is one of our latest flowering plants often surviving well into November. (I have also seen it flowering in December!) The daisy-like flowers have a central yellow disc and white strap-like rays, and grow to 5 cm across; leaves are finely dissected and thread-like.

It is found in waste places and roadsides in the late summer to late fall.

A similar species, Mayweed or Stinking Chamomile (*Anthemis cotula*) is smaller, with less dissected leaves, and flowers less than 2.5 cm across, and the foliage gives off a most unpleasant odour!

The Ox-eye Daisy has simple leaves and the central flower disc is depressed.

Plants and Wildflowers of Newfoundland

YARROW
(Milfoil)
Achillea millefolium
Daisy family

The Yarrow is an erect herb whose height varies greatly with environment, from a few centimeters on frequently mown lawns and fields to 1 m in undisturbed areas. After the Common Dandelion this species is probably the most common weed in Canada. It is aromatic, and soft, with much dissected fern-like leaves. Flat compact flowerheads are composed of five-rayed flowers each resembling a five-petalled flower. The rays are usually white but are often pinkish in colour or even light mauve.

It is found on lawns, fields, roadsides and waste places in the summer to late fall.

Michael Collins

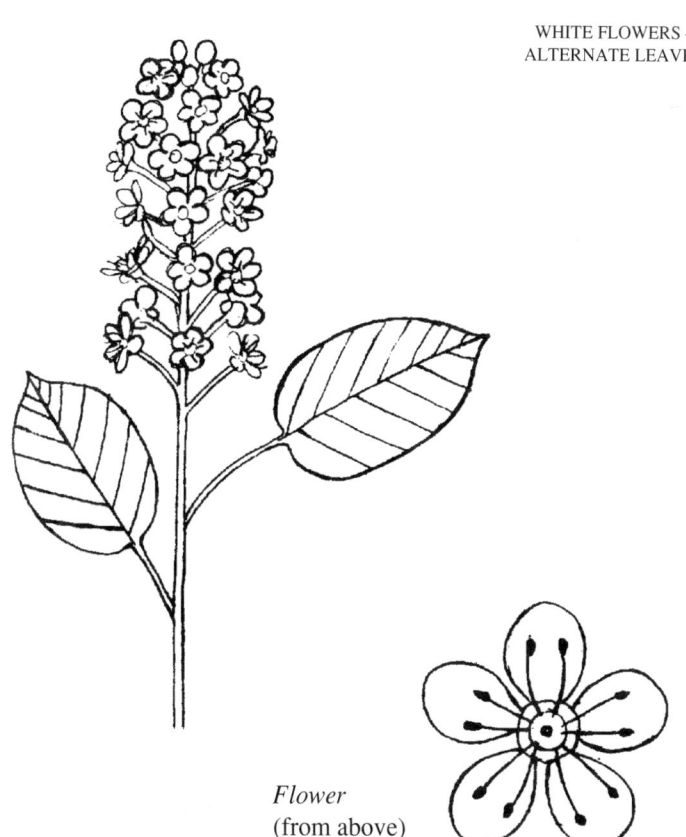

Flower
(from above)

CHOKE CHERRY
Prunus virginiana
Rose family

The Choke Cherry is a small tree with pointed ovate leaves, with stalks almost as long as the blades. It has dense, terminal, erect, cylindrical clusters of five-petalled flowers and produces hanging clusters of stalked red berries which are edible, but have a 'choking' taste.

It is found on riverbanks, damp areas, forest margins and roadsides during early summer.

(Pin Cherry does not have terminal flower clusters.)

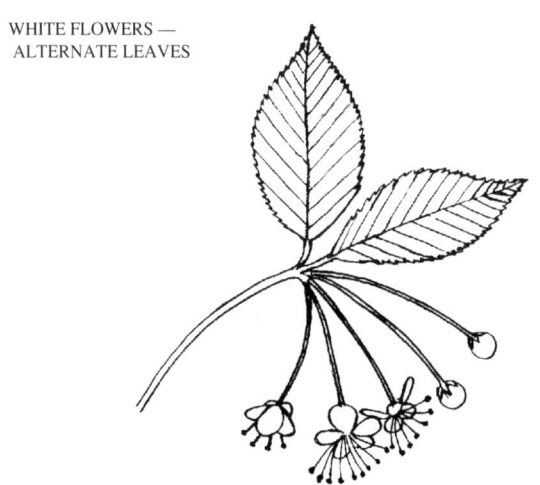

Flower
(from above)

PIN CHERRY
(Bird Cherry, Fire Cherry)
Prunus pennsylvanica
Rose family

This is a very common small tree with ovate leaves which are pointed with small teeth. It bears clusters of five-petalled flowers on long stalks and produces edible red berries on long stalks.

It is found in fields, forest clearings, open woods, burned-over areas, and roadsides in the spring.

(Choke Cherry has terminal flower clusters.)

52   Michael Collins

CHUCKLEY PEAR
(Wild Pear, Indian Pear, Juneberry, Service Berry,
Shadbush, Saskatoon)
Amelanchier species
Rose family

Chuckley Pears range from shrubs to small trees. The leaves
are alternate, ovate to oblong, toothed with pointed tips; often
with a purplish tinge. The five-petalled flowers are usually in
clusters with petals longer than broad, and spread from
2 to 3 cm in diameter. Clusters of dark purple or blackish edible
berries are produced in late summer. Chuckley Pears are the first
of the tall shrubs or trees to flower in spring.

They are found in open woods, clearings, barrens, stream
banks, and forest margins in the early spring.

(At least six species are recorded for Newfoundland. Only
one, Bartram's Chuckley Pear (*Amelanchier bartramiana*) can
be easily distinguished by its solitary flowers.)

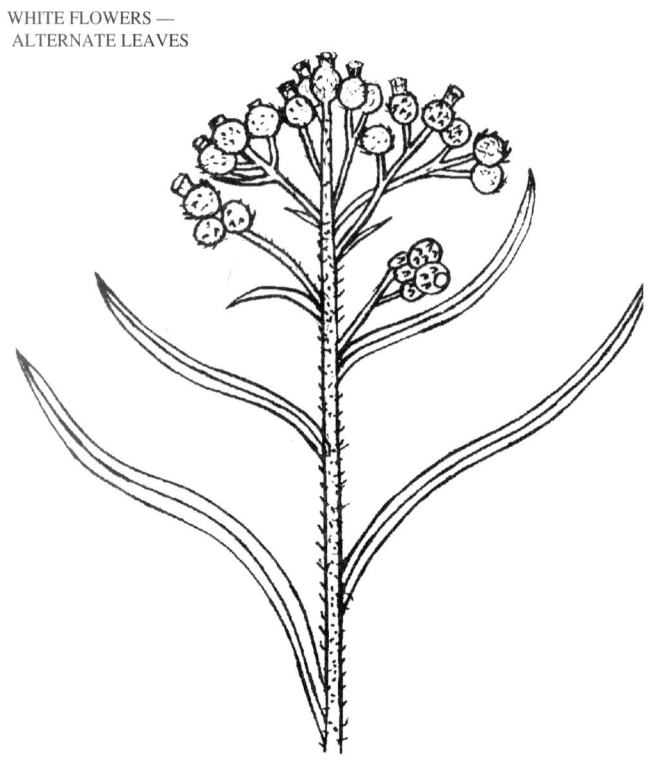

PEARLY EVERLASTING
Anaphalis margaritacea
Daisy family

The Pearly Everlasting is an erect herb which grows up to 50 cm tall. Stems and undersides of its leaves are covered with fine, white woolly down. The leaves are grass-like, dark green above and white below. The flowers of the male are small, white and globular with protruding yellow stamens in dense terminal clusters, while the female flowers are on separate plants.

It is found in waste places, roadsides and grassy areas in the summer and fall.

Michael Collins

HOODED LADY'S-TRESSES
(Hooded Ladies' Tresses)
Spiranthes romanzoffiana
Orchid family

This is an erect herb which grows to 40 cm in height. Its leaves are grass-like and narrower at the base. The flowers are small and irregular, arranged in a double spiral along the top of the stem (best observed from above) and the lower lip narrowed toward middle.

It is found in bogs and damp areas in the late summer and fall.

(Bog Candle has flowers with a long spur, and long lower lip.)

Flower
(from front)

BOG CANDLE
(Leafy White Orchis)
Platanthera dilatata   (= *Habenaria dilatata*)
Orchid family

   The Bog Candle is an erect herb which grows to 50 cm or taller with grass-like leaves. It has small and irregular flowers with tapering lower lip about the same length as the downward pointing spur.

   It is found in bogs and damp areas in the late summer and fall.

   (Hooded Lady's-tresses has flowers in double spiral.)

56   Michael Collins

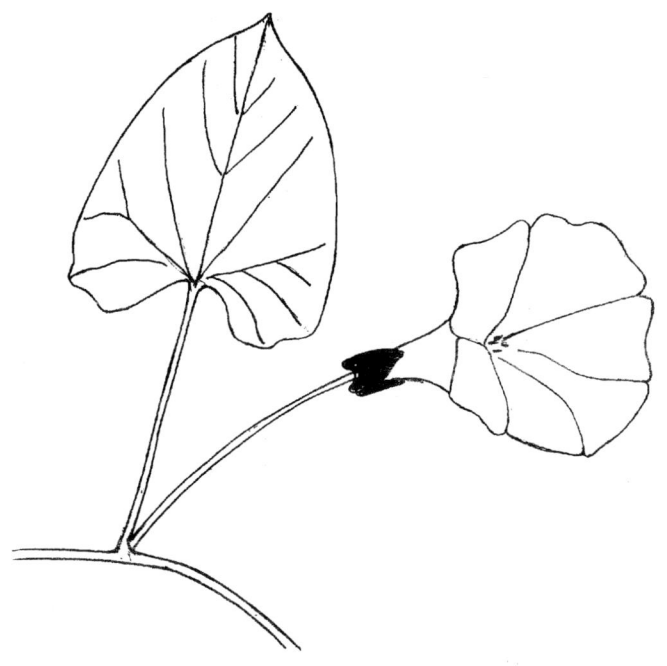

HEDGE BINDWEED
Convolvulus sepium
Morning glory family

The Hedge Bindweed is a climbing vine with weak stems which twine around other plants, fences, etc., for support. Its leaves are arrowhead-shaped on long stalks and its large white trumpet-shaped flowers spread up to 6 cm across, growing also on long stalks.

It is found in gardens and on path edges in the late summer and fall.

LEATHERLEAF
Chamaedaphne calyculata
Heath family

A low woody shrub, often in dense patches, the Leatherleaf is one of our earliest spring flowers. It has evergreen leaves, ovate and leathery to the touch. Chains of small, white, bell-like flowers are produced at the ends of horizontal branches.

It is found in bogs, pond edges and other wet areas in the early spring.

Michael Collins

BLUEBERRY
Vaccinium angustifolium
Heath family

This is an erect, low growing woody shrub which reaches up to 50 cm in height. The younger twigs are green, and the older ones are reddish in colour. Its elliptical leaves are bright green above and paler below, and its flowers hang in bell-like clusters at the stem tips. Edible, sweet, dark blue spherical berries are produced in late summer.

It is found in open barrens, peatlands, and open areas of forests in the early summer.

*Individual
Flower*

WILD LILY-OF-THE-VALLEY
(Canada Mayflower)
Maianthemum canadense
Lily family

The Wild Lily-of-the-Valley is a small, erect, low growing herb which grows to 15 cm in height. One main stem bears both the leaves and flowers. It has two heart-shaped parallel-veined leaves with a terminal cluster of minute, four-petalled flowers. In the fall it produces a cluster of mottled red berries.

It is found on the forest floor during the spring.

Michael Collins

OX-EYE DAISY
(Oxeye daisy)
Chrysanthemum leucanthemum
Daisy family

An erect herb which grows to 60 cm. The leaves of the Ox-eye Daisy are small, narrow and lobed. It is our typical daisy with large flowers which spread to 5 cm in diameter with white petal-like rays that radiate out from a central yellow disc with a slight depression.

It is found in fields, roadsides and waste places in the summer and fall.

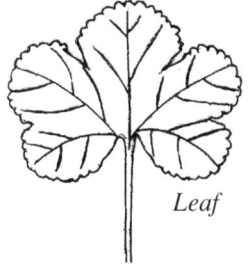

Leaf

Flower
(from above)

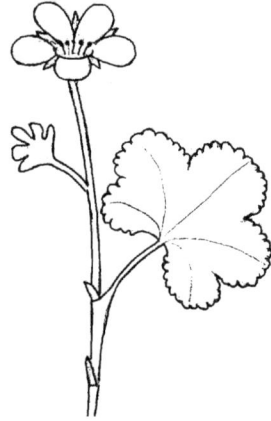

BAKEAPPLE
(Cloudberry, Baked-apple Berry)
Rubus chamaemorus
Rose family

The Bakeapple is a low growing herb, reaching a height of up to 15 cm, with usually two five-lobed leaves and one single five-petalled white flower that spreads to about 15 mm in diameter. It produces orange coloured blackberry-like fruit.

It is found in bogs and peaty places in the early summer.

Michael Collins

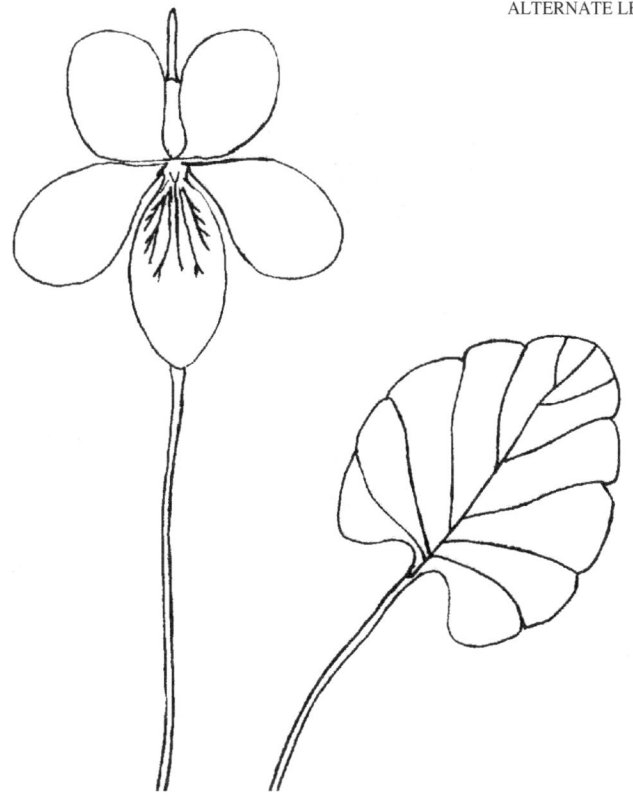

NORTHERN WHITE VIOLET
Viola pallens
Violet family

This is a low-growing herb, and has flowers and leaves on separate stalks. The nearly-round leaves grow on long (up to 7 cm), slender stalks. Its flowers are irregular, five-petalled with purple guide lines on lower lip which spread to 13 mm across, and are held on long erect stalks.

It is found in damp areas in the spring.

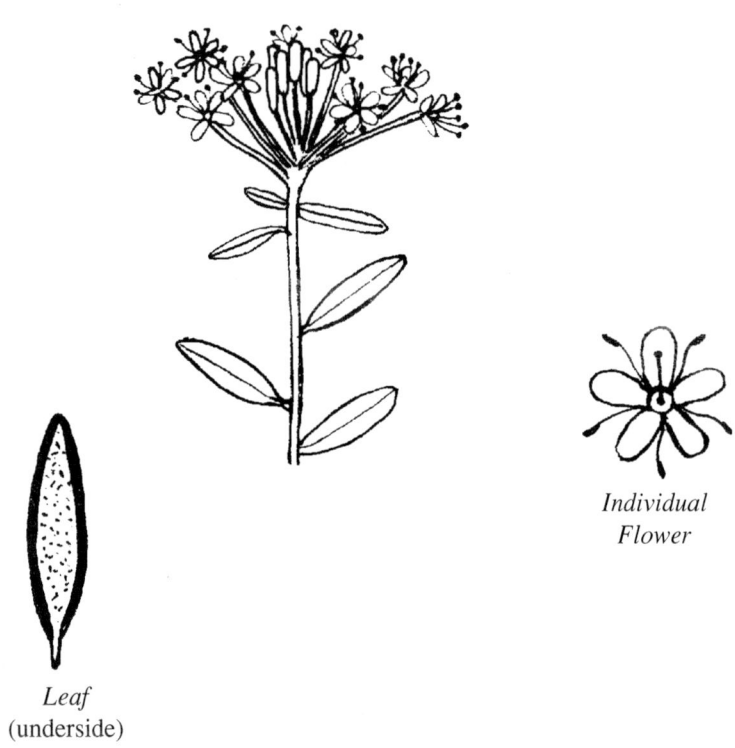

Individual
Flower

Leaf
(underside)

LABRADOR TEA
Ledum groenlandicum
Heath family

The Labrador Tea is an erect, branching, evergreen shrub growing to 1 m high. It has elliptical evergreen leaves that are green above and covered with orange fuzz below. Its five-petalled flowers grow in compact terminal heads.

It is found in bogs and peaty areas during the early summer.

Michael Collins

MEADOWSWEET
Spiraea latifolia
Rose family

An erect, branching, wiry shrub, the Meadowsweet grows up to 1.5 m in height. Its alternate leaves are ovate and toothed, and are closely set on the stem. Spikes of small five-petalled white flowers, sometimes pink, grow at the tips of the stem.

It is found on stream banks, damp areas and roadsides in the summer and early fall.

Plants and Wildflowers of Newfoundland 65

PURPLE CHOKEBERRY
Aronia prunifolia (= *Pyrus floribunda*)
Rose family

The Purple Chokeberry is an erect, much branched shrub, reaching a height of up to 1 m, with woolly twigs and flower stalks. It has elliptical leaves, with pointed tips; finely toothed; and white woolly beneath. It produces terminal clusters of small five-petalled white flowers, sometimes with a pinkish tinge, and edible purplish-black berries.

It is found in bogs, heaths, and damp areas in the summer.

Michael Collins

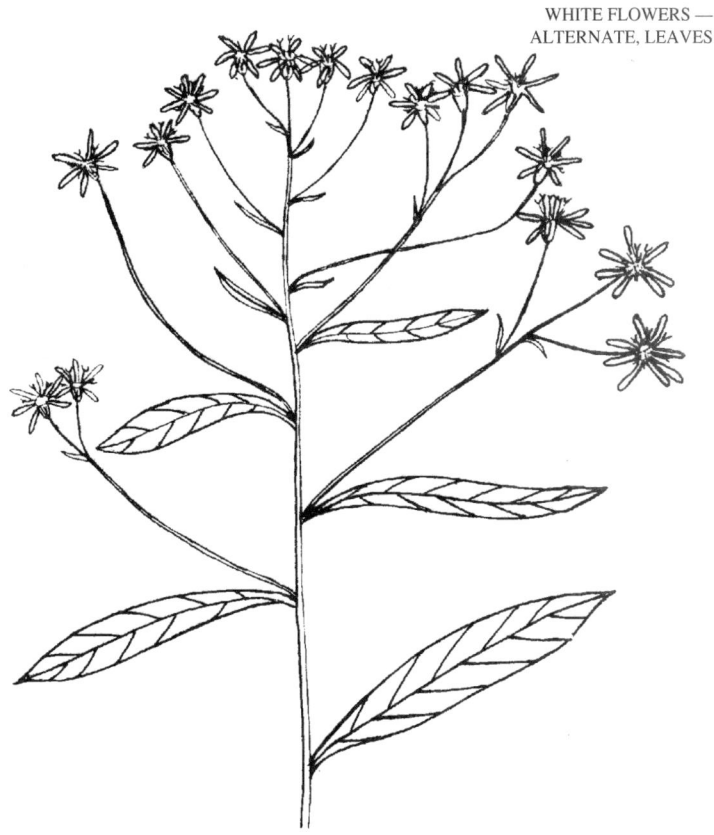

FLAT-TOPPED WHITE ASTER
Aster umbellatus
Daisy family

This is a tall, erect herb which grows to 2 m high. It has long, narrow lance-shaped toothless leaves and a flat-topped cluster of flowers. Each flower has fewer than ten rays spreading up to 15 mm across with yellow discs.

It is found on pond edges and damp areas in the late summer and fall.

(Panicled Aster (*Aster simplex*) is similar but flowers have more than 20 rays, and not in flat-topped clusters.)

Yellow Flowers

No leaves present at flowering time

COLTSFOOT
Tussilago farfara
Daisy family

The Colstfoot is an erect herb that grows to 15 cm tall with flowers that appear before the leaves. Its stalk has red-tinged scales. Large horseshoe-shaped leaves appear after the flower. The single yellow dandelion-type flower at the top of a solid stalk spreads to 25 mm across.

It is found in open waste areas, and on roadsides in the early spring.

(Common Dandelion has a hollow stalk, and leaves at flowering time.)

Michael Collins

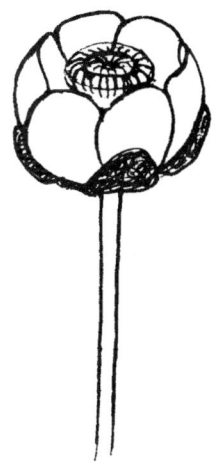

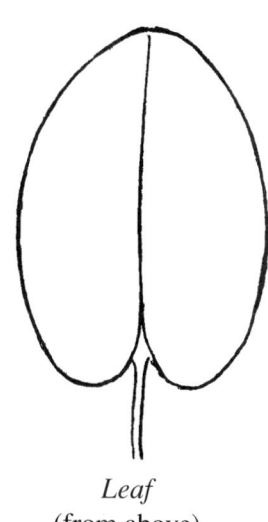

Leaf
(from above)

BULLHEAD LILY
(Yellow Pond Lily, Cow Lily, Spatterdock)
Nuphar variegatum
Water lily family

The floating leaves and flowers of the Bullhead Lily are connected by long stalks to an underground stem in the pond bottom. Its large, ovate and green leaves, with rounded bases, float
on the water surface. Large almost spherical yellow flowers are
held above the water surface on stout stalks. The yellow 'petals'
are actually sepals and each flower opens for just one day.

It is found in shallow ponds and quiet stretches of rivers during the summer

Leaves in Basal Rosette

COMMON DANDELION
(Piss-a-beds, Dumbledor, Faceclock)
Taraxacum officinalis
Daisy family

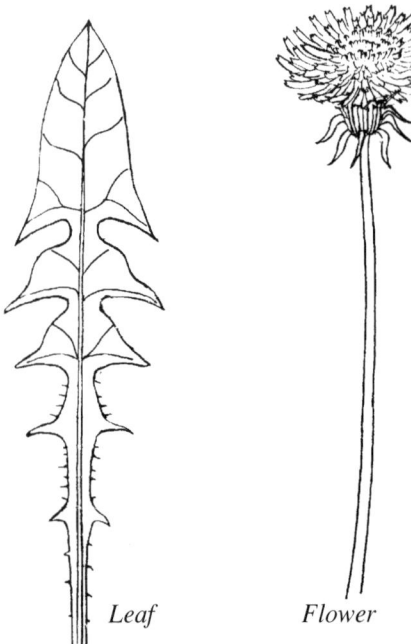

Leaf   *Flower*

The Common Dandelion is an herb with leaves in a basal rosette and a single flower on an erect stalk which grows up to 50 cm in height. It has a basal rosette of large, toothed leaves. The single flower, with many golden-yellow rays, spreads to 3 cm across at the top of a single, hollow, latex-filled stalk, and flower bracts point downward. It is our most common weed and most recognizable flower. The appearance of its flowers early in the year are a sure sign of spring! The Common Dandelion, although the bane of gardeners, can be a very useful plant. The leaves can be used as spring greens, the flowers can be made into wine, and the underground rootstock can be roasted and used to make a coffee!

It grows on lawns, in fields, on roadsides, and in waste places in the spring, but often flowers again in the late summer and fall.

(Coltsfoot has a solid stalk but no leaves at the time of flowering. Fall Dandelion has several flowers and a solid stalk.)

Michael Collins

FALL DANDELION
Leontodon autumnalis
Daisy family

The Fall Dandelion is an herb, which grows to 50 cm tall, has a basal rosette of leaves and several flowers arising from branched stalks. This is one of our hardiest plants with blooms that can survive frosts and light snowfalls. Its leaves, in a basal rosette, have tooth-like projections. Numbers of dandelion-like flowers arise from a branched, thin, wiry, solid stalk which bears small scales. Just as the appearance of the flowers of the Common Dandelion heralds the arrival of spring, the disappearance of the blooms of the Fall Dandelion signals the onset of winter!

It is found on lawns, roadsides, and in waste places in the late summer and fall. (Also, the Fall Dandelion can often be found flowering through late November or even into early December in the St. John's area).

(Common Dandelion has hollow stalk with a single flower.)

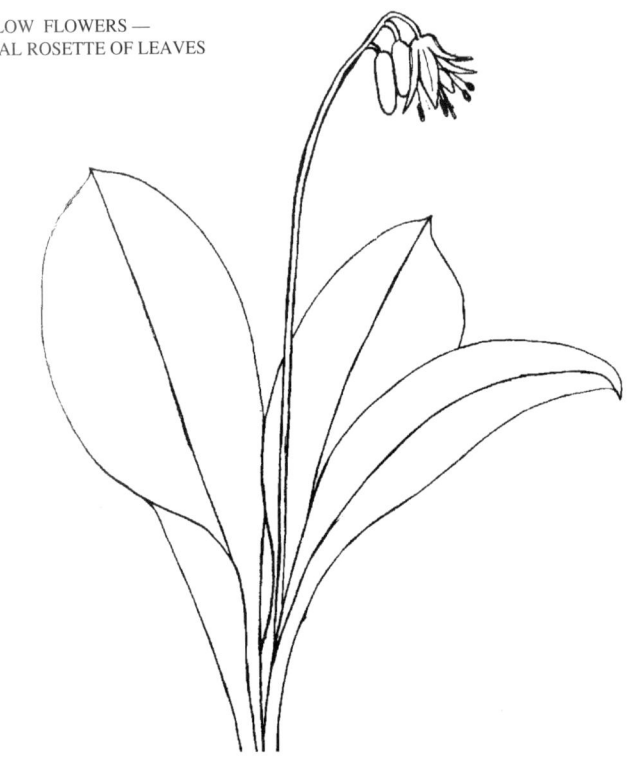

CLINTONIA
(Corn Lily, Poison Berry, Bluebead)
Clintonia borealis
Lily family

This is an erect herb with basal leaves which grows up to 50 cm in height. It has three prominent parallel-veined, oval leaves arising from the base of the stem. An erect stalk bears several hanging, bell-like flowers, each with six petal-like parts, and it produces a cluster of dark blue berries at the tip of the stalk which are sometimes mistaken for blueberries.

It is found in dark woods and shaded locations in the late spring and early summer.

Michael Collins

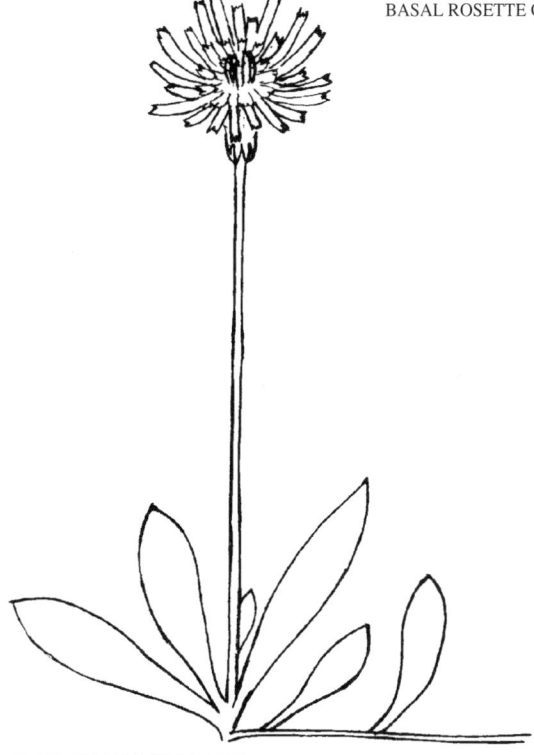

YELLOW FLOWERS —
BASAL ROSETTE OF LEAVES

MOUSE-EAR HAWKWEED
Hieracium pilosella
Daisy family

The Mouse-ear Hawkweed is an herb and has a basal rosette of leaves and a vertical unbranched flower stalk which grows up to 30 cm tall. It spreads by overground runners and can form extensive mats. It has a basal rosette of elliptical leaves that are white woolly beneath. It has a single dandelion-like flower at the tip of the stalk.

It is found in grassy areas, roadsides and waste places in the summer and early fall.

(Other hawkweeds do not have solitary blossoms.)

Plants and Wildflowers of Newfoundland   73

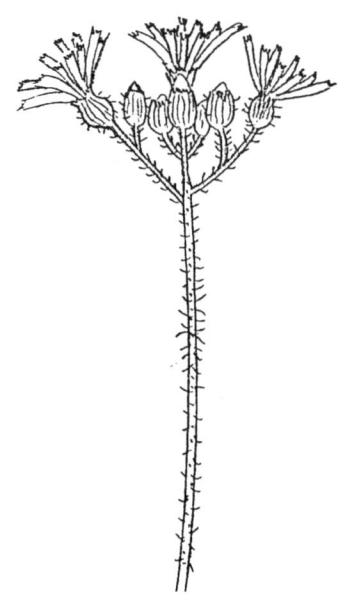

Basal Leaf

KING DEVIL
(Field Hawkweed)
Hieracium pratense
Daisy family

An herb with a basal rosette of leaves, and multiple flowers arising from a single erect stem, the King Devil's stem and leaves are very hairy. It spreads by runners and grows up to 75 cm tall. Its basal rosette of long, narrow, pointed leaves are covered with black hairs and the single erect stem bears clusters of dandelion-like flowers at the tip.

It is found in fields and on roadsides in the summer.

(Mouse-ear Hawkweed has a single blossom; Common Hawkweed has leaves along the whole length of the stem; Canada Hawkweed does not have a basal rosette.)

74   Michael Collins

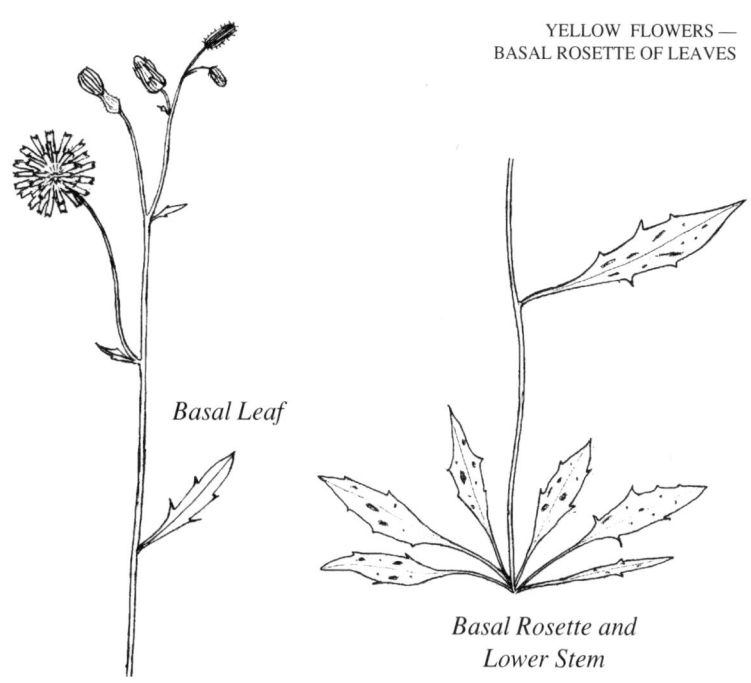

Basal Leaf

*Basal Rosette and
Lower Stem*

COMMON HAWKWEED
Hieracium vulgatum
Daisy family

An erect herb with a basal rosette of leaves, and a few leaves further up the stem, the Common Hawkweed grows up to 1 m tall. In addition to a basal rosette, there are several leaves on the upper stem. The basal leaves are slightly toothed, pointed, ovate with stalks and are mottled with purple. Upper stem leaves are usually smaller and sharp pointed and stalkless, with a few prominent teeth. Several large (up to 4 cm) dandelion-like flowers on separate stalks grow at the tip of the stem.

It is found on roadsides, path edges, and waste ground in the summer and fall.

(Mouse-ear Hawkweed has solitary blossoms; King Devil has leaves in a basal rosette; Canada Hawkweed has no basal rosette.)

Opposite Leaves

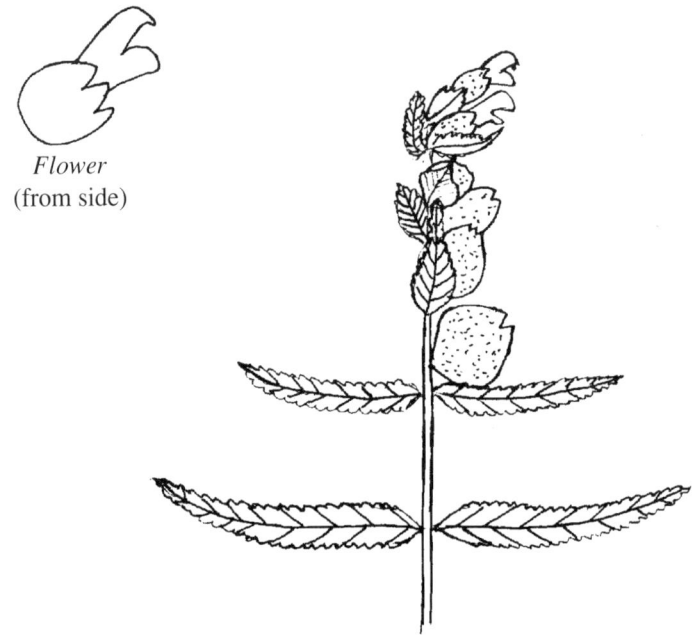

Flower
(from side)

YELLOW RATTLE
Rhinanthus crista-galli
Figwort family

An erect herb which grows to 25 cm, the Yellow Rattle has opposite leaves that are long, narrow and toothed. Yellow petals project from within a sac-like calyx. After the flower is fertilized, the calyx inflates, and the seeds rattle around in this outer casing. It is partially parasitic on grass roots.

It is found in fields and on roadsides in the summer.

Michael Collins

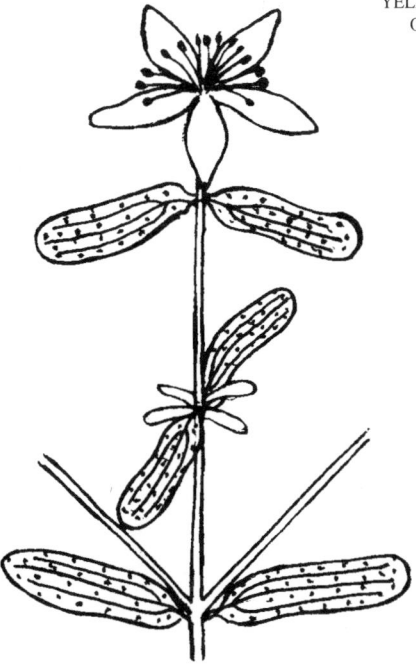

COMMON ST. JOHN'S-WORT
(Common St. John's-Wort)
Hypericum perforatum
St. John's-wort family

The Common St. John's-wort is a much branched, erect herb which grows to 75 cm tall, with rounded rectangular opposite leaves. When held up to the light, the leaves show dozens of translucent spots. Numbers of bright yellow, five-petalled flowers grow at the tips of stalks and are up to 3 cm in diameter.

It is found in fields, on roadsides, and in waste places in the late summer and fall.

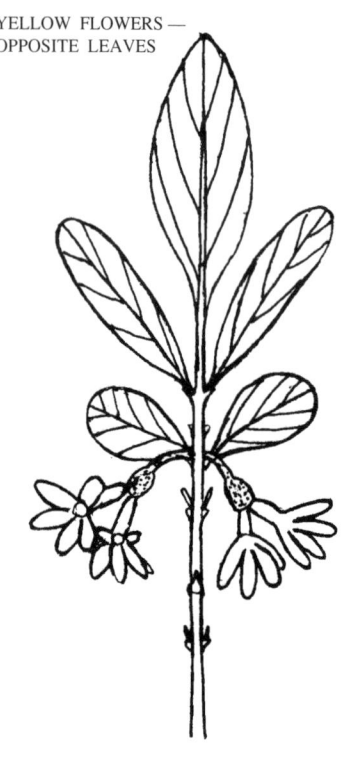

YELLOW FLOWERS —
OPPOSITE LEAVES

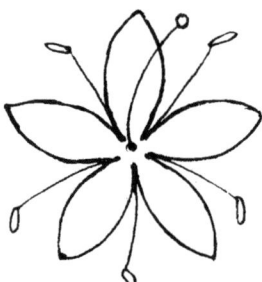

NORTHERN HONEYSUCKLE
(Northern Fly-honeysuckle)
Lonicera villosa
Honeysuckle family

Northern Honeysuckle is a small erect shrub that grows up to 1 m in height. It has opposite leaves, that are roughly ovate. Flowers, light yellow, tubular and five-lobed, arise in pairs from a central base and spread to 12 mm across. After fertilization the ovaries of each flower pair fuse to form an edible, blue, double-seeded berry.

It is found in damp areas and bogs in the late spring.

Michael Collins

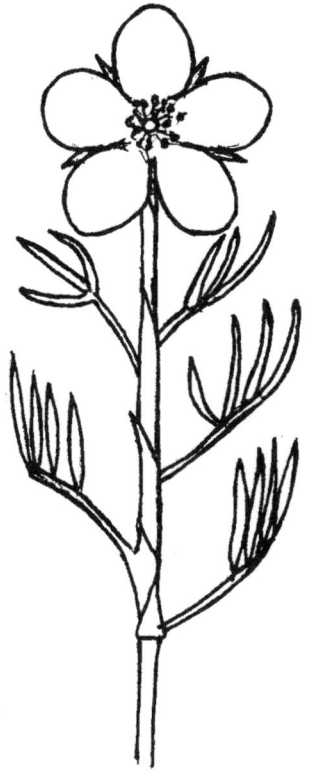

SHRUBBY CINQUEFOIL
Potentilla fruticosa
Rose family

This is an erect woody shrub that grows up to 1 m high. Its leaves are compound with 5 or more toothless leaflets and it bears clusters of five-petalled yellow flowers with sepals as long as the petals.

It is found in bogs and damp areas in the late summer and fall.

Flower
(from above)

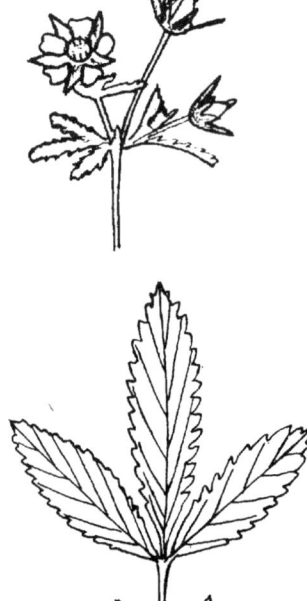

ROUGH CINQUEFOIL
Potentilla norvegica
Rose family

The Rough Cinquefoil is an erect branched herb that grows up to 50 cm and has hairy stems. Its leaves and flowers arise from the same stalk. The leaves are compound with three, toothed, elliptical leaflets and two large stipules at the leaf base. It bears clusters of five-petalled yellow flowers with sepals projecting beyond the petals.

It is found on roadsides and in waste places in the summer and fall.

Michael Collins

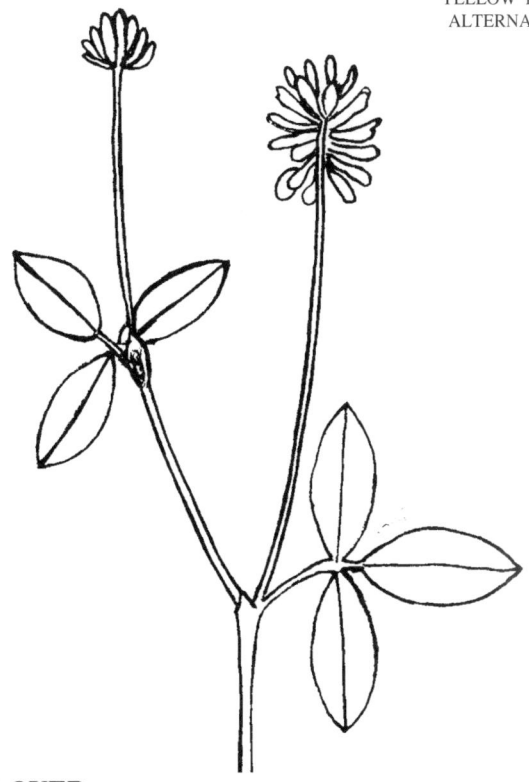

HOP CLOVER
Trifolium agrarium
Rose family

This is a low growing, erect herb, reaching a height of up to 25 cm, with leaves and flowers that arise from the same stalk. Its leaves are compound with three pointed, ovate leaflets, and it has compact clover-like flowerheads. When the flowerheads wither the individual flowers turn down and become brown, and then resemble dried hops, hence the plant's common name.

It is found in grassy areas, on roadsides, and in waste places in the summer and fall.

Flowers
(seen from above)

BIRDFOOT TREFOIL
(Birdsfoot Trefoil)
Lotus corniculatus
Pea family

The Birdfoot Trefoil is an erect or prostrate herb, reaching up to 30 cm in height, with leaves and flowers arising from same stalk. Leaves are compound with five leaflets—three clover-like leaflets and two more at the base resembling stipules. Distinctive yellow or orangey pea-like flowers grow in clusters of three or more. When the seed pods ripen the triplet of pods resembles a bird's foot, hence the common name.

It is found in grassy areas, on roadsides, and in waste ground in the summer.

Michael Collins

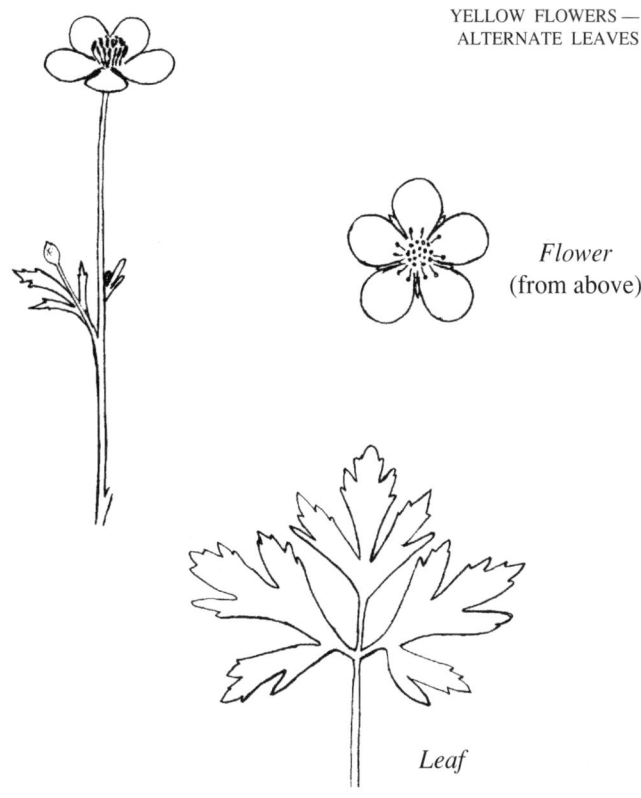

Flower
(from above)

Leaf

CREEPING BUTTERCUP
Ranunculus repens
Buttercup family

A creeping herb which spreads by overground runners, the Creeping Buttercup grows to a height of 30 cm. After the Common Dandelion this is the lawn owner's next most common weed with which to contend. It has compound leaves with three stalked, divided leaflets, often with dark blotches, and familiar yellow five-petalled flowers.

It is found on lawns, gardens, meadows, roadsides, and in waste places during the summer and fall.

(Common Buttercup has 5 or more leaflets.)

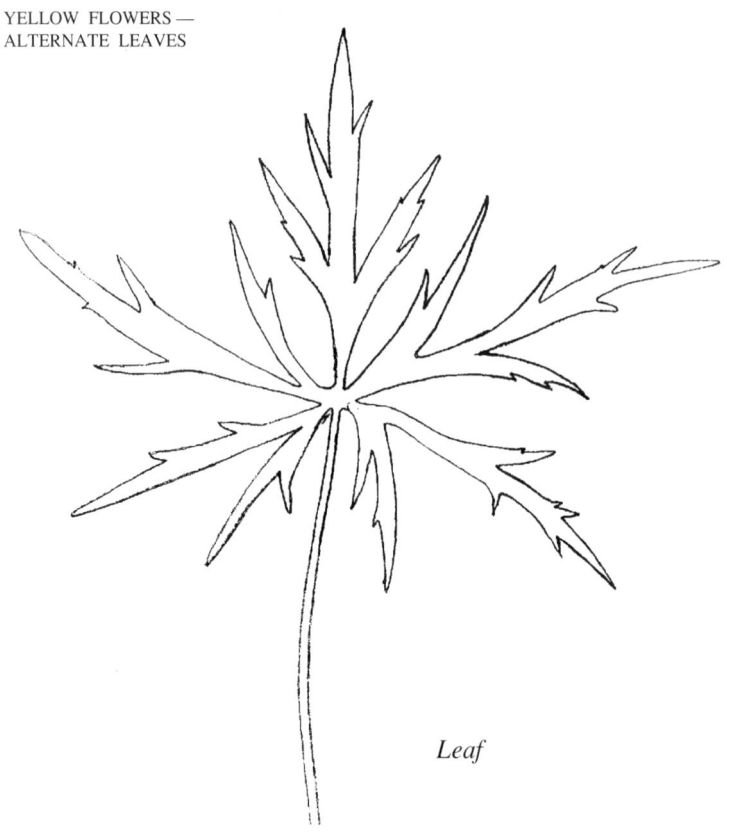

Leaf

COMMON BUTTERCUP
(Tall Buttercup)
Ranunculus acris
Buttercup family

Our familiar buttercup, the Common Buttercup grows up to 1 m tall. Its leaves are compound with five to seven stalkless, divided leaflets. It produces the familiar five-petalled, bright yellow flowers.

It is found in fields, meadows, in ditches and damper areas during the summer to fall.

(Creeping Buttercup has 3 leaflets.)

Michael Collins

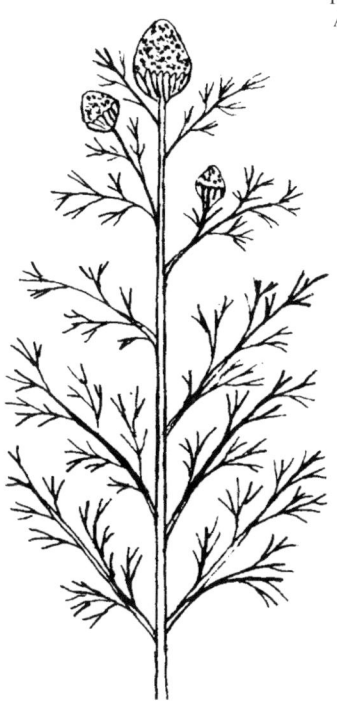

PINEAPPLEWEED
(Pineapple Weed)
Matricaria matricarioides
Daisy family

A low growing, erect herb, the Pineappleweed grows to a height of 25 cm. Its leaves are finely divided into many narrow segments and give off a pineapple smell when bruised, hence the common name. The flowers are almost spherical and yellow, lacking the showy rays of other daisy-like plants. This plant resembles a small Scentless Chamomile (to which it is related) whose flowers have lost their white petal-like rays.

It is found in waste places, roadsides, paths, playing fields, and even cracks in sidewalks, in the summer to late fall.

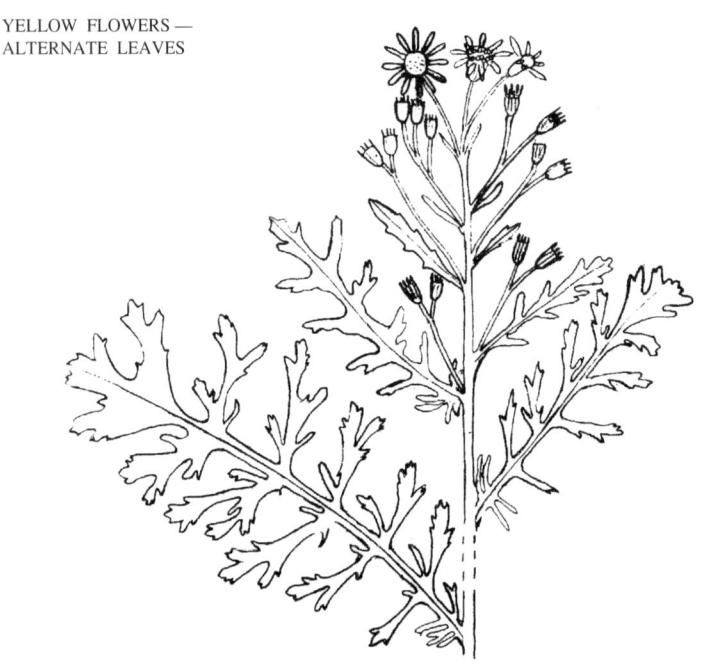

TANSY RAGWORT
(Common Ragwort)
Senecio jacobaea
Daisy family

The Tansy Ragwort is a tall erect herb that grows up to 1.5 m, and is often found in dense stands. Its leaves are large and finely dissected, and it produces flat-topped clusters of small (2 cm) yellow flowers at the top of the stem, each with usually ten or more yellow rays. It is one of our tallest and most distinctive wildflowers, as well as being one of our most hardy. It can survive light frosts and snowfalls, and its flowers are often among the last of the year. In mild years its flowers can last into December in the St. John's area.

It is found in fields, on roadsides, and in waste places.

Michael Collins

LANCE-LEAVED GOLDENROD

Solidago graminifolia   (= *Euthamia graminifoli*a)
Daisy family

A tall, erect herb, the Lance-leaved Goldenrod grows up to
1.5 m in height. The stem is crowded with many long, slender,
grass-like leaves, the lower leaves showing three parallel veins.
Many clusters of minute yellow flowers grow at the tip of the
stem.

It is found on riverbanks, in ditches and other damp areas in
the late summer and fall.

BUTTER-AND-EGGS
(Toadflax)
Linaria vulgaris
Snapdragon family

This is an erect herb that grows up to 75 cm in height. Its stem is crowded with many linear, grass-like leaves. It produces a terminal spike of many yellow and orange snapdragon-like flowers, each with a yellow upper and orange lower lip, and a long spur.

It is found in gardens, open areas, on roadsides and in waste ground in the summer and fall.

(Blue Toadflax has light mauve flowers.)

Michael Collins

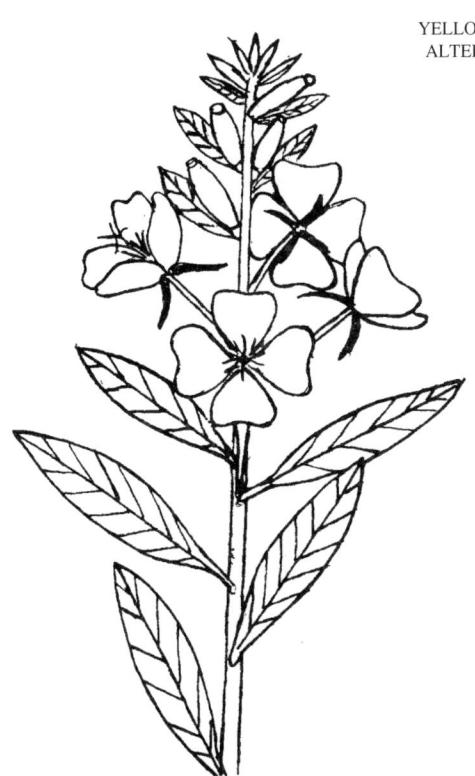

COMMON EVENING-PRIMROSE
(Common Evening Primrose)
Oenothera biennis
Evening-primrose family

A tall, erect herb, the Common Evening-primrose grows up to 1 m in height. Its stem is crowded with stalked, elliptical leaves. Many long-stalked, four-petalled flowers grow towards the top of the stem, each with a distinctive X-shaped stigma, and reflexed petals that grow up to 5 cm in diameter. Its flowers fully open towards sunset. After fertilization, the flowers produce distinctive urn-shaped seed receptacles.

It is found on roadsides and waste places in the summer.

(Small-flowered Evening-primrose has 2 cm flowers.)

SMALL-FLOWERED EVENING-PRIMROSE
Oenothera parviflora
Evening-primrose family

This is a tall, erect herb, growing to a height of 50 cm, with a stem not as crowded with elliptical leaves as the Common Evening-primrose. Its many long-stalked, four-petalled flowers with X-shaped stigmas, and reflexed petals grow up to 2 cm in diameter, and its flower stalks have swollen bases. After fertilization the flowers produce distinctive urn-shaped seed receptacles on long stalks.

It is found on roadsides and in waste places in the late summer and fall.

(Common Evening-primrose has larger, 5 cm flowers.)

Michael Collins

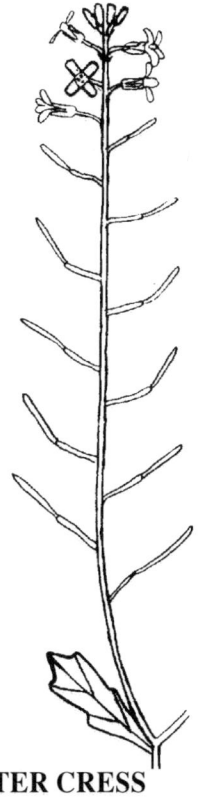

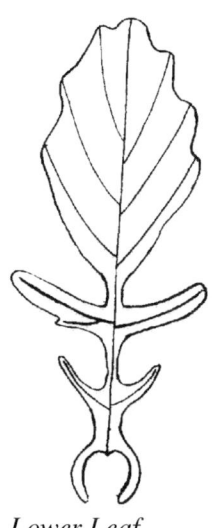

Lower Leaf

WINTER CRESS
(Yellow Rocket, Herb Barbara)
Barbarea vulgaris
Mustard family

The Winter Cress is an erect, much branched herb that grows up to 60 cm in height. Its upper leaves are lobed, with lower leaves showing very pronounced lobes, and two basal lobes which curve downwards. Many small (7 mm) four-petalled yellow flowers grow on stalks at right angles to the stem, and long, narrow seed pods develop from the flowers.

It is found in meadows, gardens and waste places in the spring to fall.

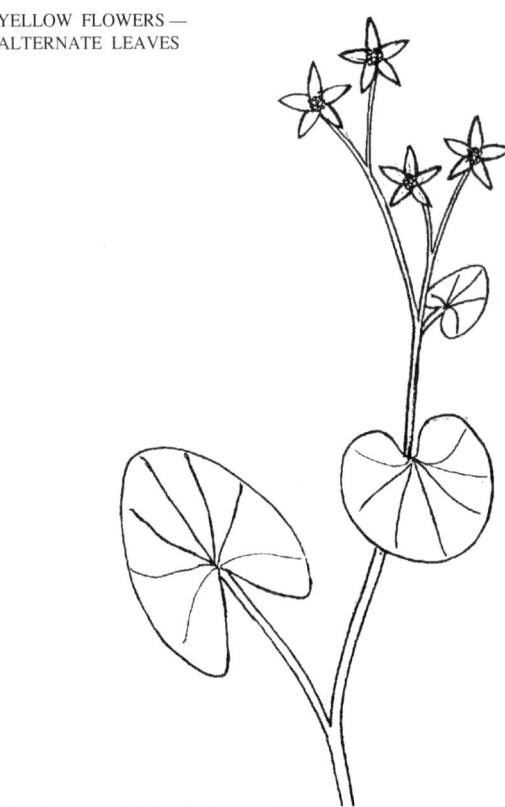

MARSH MARIGOLD
(Cowslip)
Caltha palustris
Buttercup family

This is a tall, erect herb of standing water which grows up to 50 cm in height. Its lower leaves are kidney-shaped with long stalks, and upper leaves less rounded and with short stalks. Large, bright yellow buttercup-type flowers, with five or more petal-like sepals are produced, which spread up to 25 mm across.

It is found in ditches and other areas of standing water in the spring.

(Buttercups are smaller with compound leaves.)

Michael Collins

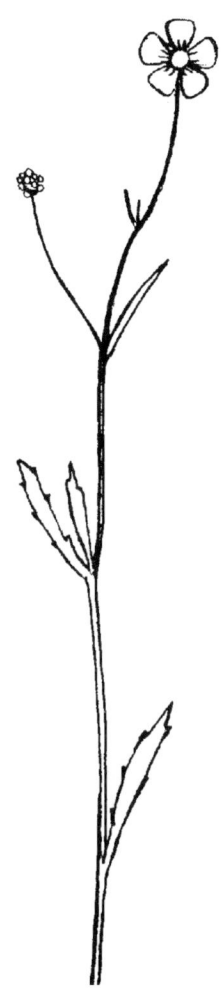

SPEARWORT
Ranunculus flammula
Buttercup family

The Spearwort is the low straggling buttercup of muddy areas which grows up to 20 cm tall. Leaves are lance-shaped with several teeth, and its flowers, five-petalled, spread up to 11 mm across.

It is found in streams, pond banks, and other damp, muddy places in late summer and fall.

*Individual
Flowerhead*

BOG GOLDENROD
Solidago uliginosa
Daisy family

The Bog Goldenrod is an erect herb of bogs which grows to 1 m tall. A single stem supports many close-set, slightly toothed, lance-shaped leaves whose bases clasp the stem. It has many small flowers in dense clusters at the top of the stem.

It is found in bogs and other damp areas during the late summer and early fall.

94   Michael Collins

ROUGH-STEMMED GOLDENROD
Solidago rugosa
Daisy family

This is a tall, erect herb which grows to 1.5 m in height. It is our commonest goldenrod, and often found in extensive stands. Leaves are shield-shaped and toothed, each with a prominent central vein. It bears many side branches near the top of the stem with many small yellow flowerheads.

It is found in fields, roadsides and waste places in the late summer and fall.

(Canada Goldenrod has leaves with 3 prominent veins.)

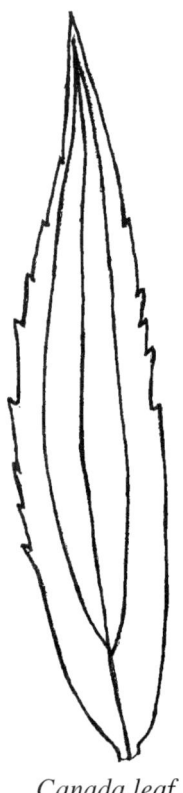

Canada leaf   *Rough-stemmed leaf*

CANADA GOLDENROD
(*Solidago canadensis*)
Daisy family

This is a very similar plant to Rough-stemmed Goldenrod which grows in similar habitats. It can be distinguished from the Rough-stemmed by its leaves which have three prominent parallel veins.

(Bog Goldenrod is smaller with leaf bases which clasp the stem.)

Michael Collins

LARGE-LEAVED GOLDENROD
Solidago macrophylla
Daisy family

An erect herb of shaded places which grows to 1 m in height, the Large-leaved Goldenrod has heart-shaped lower leaves on long stalks. Its individual flowers are larger (up to 15 mm across) than in other local goldenrods.

It is found in woods, and on path edges in the summer and early fall.

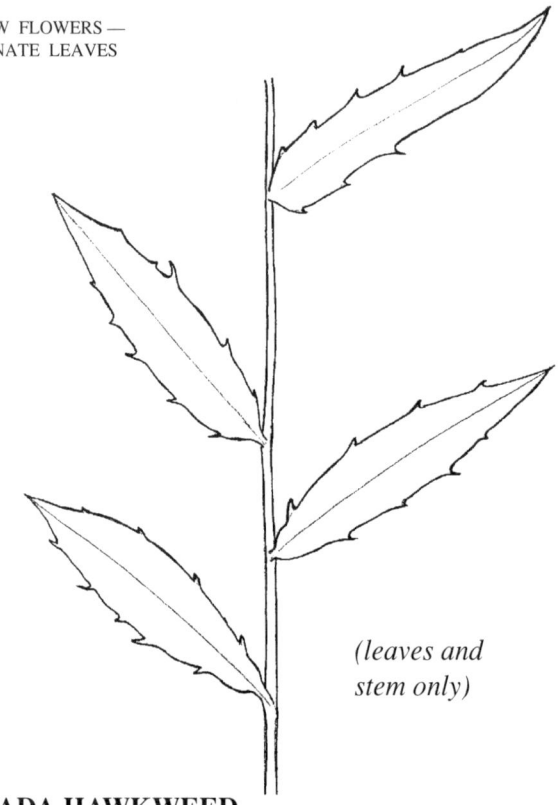

*(leaves and
stem only)*

CANADA HAWKWEED
Hieracium canadense
Daisy family

This is an erect herb which grows to 1 m in height. Toothed leaves of the Canada Hawkweed occur along the whole length of the stem with several large (2.5 cm) dandelion-like flowers on separate stalks at the tip of the stem.

Most other local hawkweeds have basal rosettes of leaves but no leaves on the stem.

It is found on roadsides, and path edges during the summer and fall.

(Common Hawkweed has a basal rosette and a few stem leaves.)

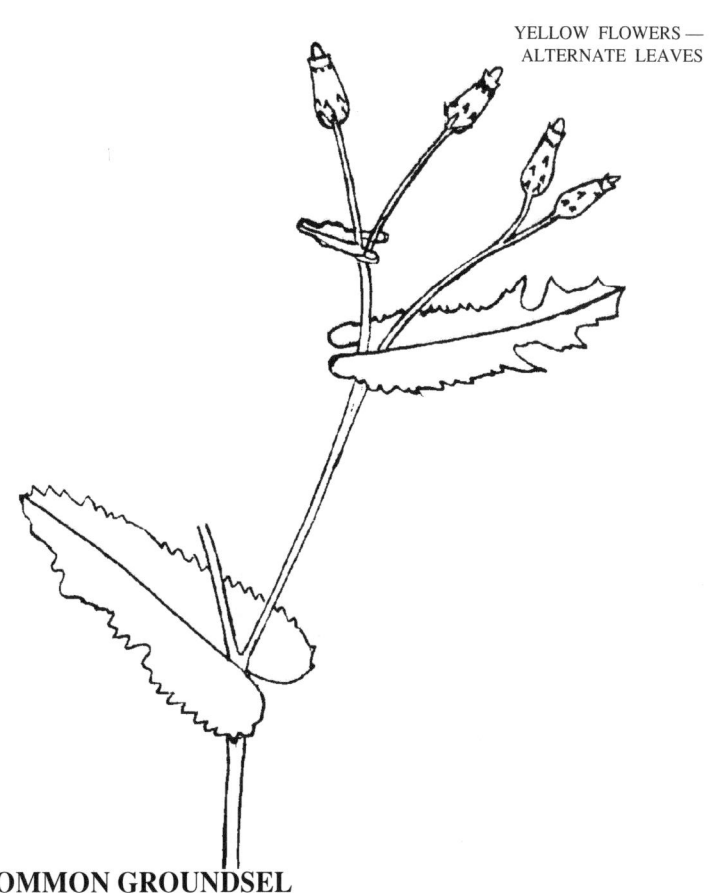

COMMON GROUNDSEL
Senecio vulgaris
Daisy family

The Common Groundsel is an erect herb which grows to 50 cm in height. Its leaves are irregularly toothed with the leaf bases extending back beyond the stem. Flowers are numerous but small, with no yellow rays, and the flower bases are covered with black-tipped bracts.

It is found in gardens and waste ground in the summer to fall.

(Sticky Groundsel has conspicuous flower rays.)

Plants and Wildflowers of Newfoundland 99

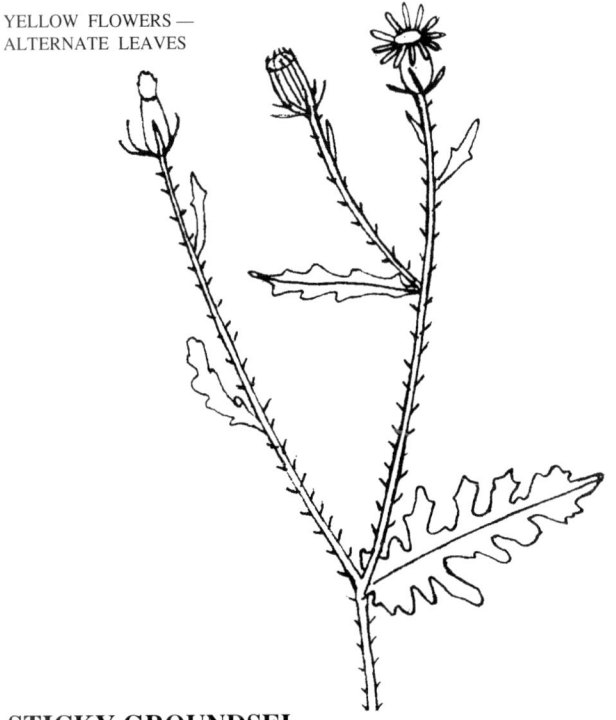

STICKY GROUNDSEL
(Stinking Willie, Stinking Groundsel)
Senecio viscosus
Daisy family

This is an erect herb which grows to 30 cm in height. (In 'ideal' conditions it can be shrub-like and 1 m tall.) The Sticky Groundsel is covered with sticky hairs, often with dirt and the like stuck to them, and gives off an unpleasant odour when bruised, hence the common name. Hairs give the plant a silvery appearance. Its leaves are irregularly lobed, and flowers are numerous but small, possessing short yellow rays which spread up to 1 cm across.

It is found on waste ground in the summer and fall.

(Common Groundsel doesn't have obvious flower rays, is odourless and not sticky.)

Michael Collins

Pink and Red Flowers

Leaves in Basal Rosette

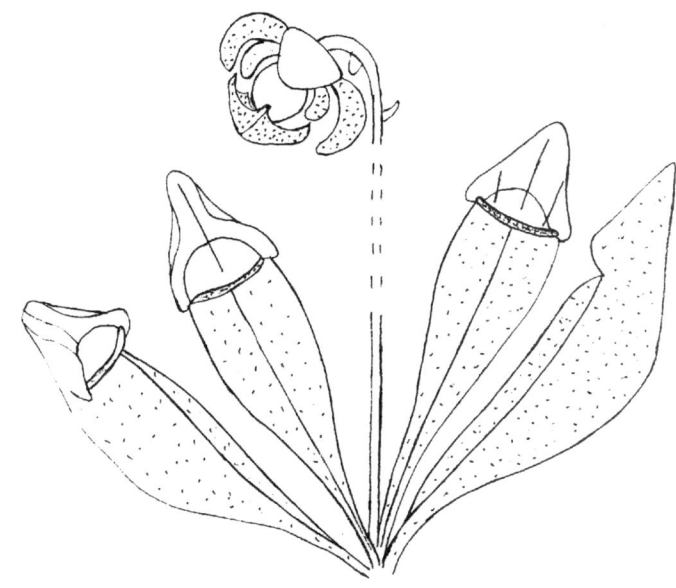

PITCHER PLANT
Sarracenia purpurea
Pitcher plant family

Our provincial flower and an unmistakeable one at that! The Pitcher Plant grows up to 60 cm tall and is insectivorous. Large hollow pitcher-like leaves arise from the base of the stalk. Each pitcher contains water in which insects drown and become digested, with the minerals being absorbed by the plant. It has a single unusual greenish-red nodding flower that has five petals and a large flattened pistil on a long, stout stalk .

It is found in wet bogs in the summer.

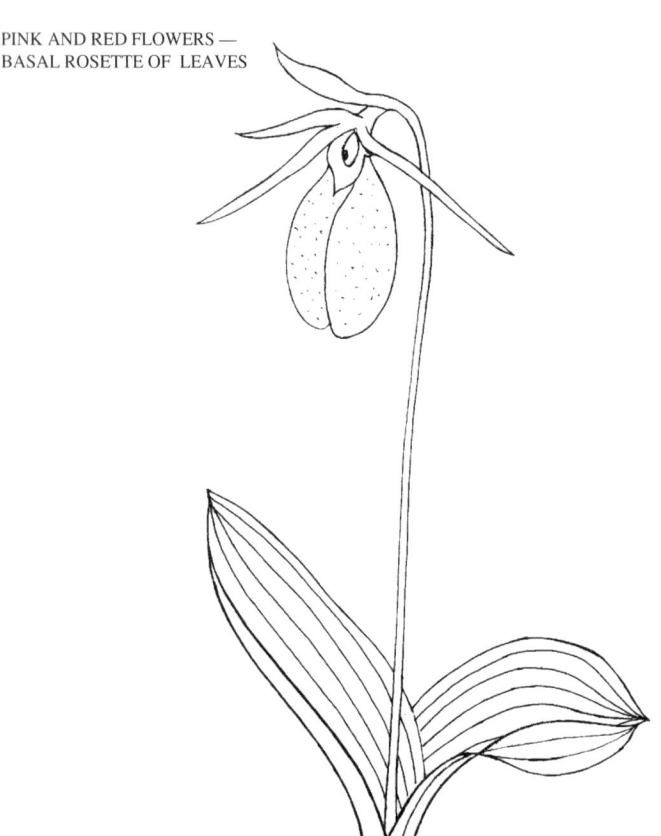

PINK LADY'S-SLIPPER
(Moccasin flower, Pink Lady's Slipper)
Cypripedium acaule
Orchid family

This is a very distinctive herb of shaded areas which grows up to 40 cm in height. It has two parallel-veined leaves that arise at ground level, and one large, bag-like pink flower at the top of the stem, although some plants may produce white flowers.

It is found in coniferous forests, and other shaded places in the spring.

Michael Collins

SHEEP LAUREL
(Lambkill, Goowitty)
Kalmia angustifolia
Heath family

This is our most common woody shrub which grows up to 1 m tall, covering vast tracts of land. Leaves are evergreen and borne in whorls of three. It has clusters of pink, saucer-shaped flowers in the leaf axils below the stem tip. Five petals are fused into a saucer-like disc. The stalked stamens lie flattened against the petals; the stigma projects outward.

It is found in barren lands, bogs, clearings in forests, and open areas during the late summer.

(Bog Laurel has opposite leaves.)

SPOTTED JOE-PYE-WEED
(Spotted Joe Pye Weed)
Eupatorium maculatum
Daisy family

A tall, erect herb of wet situations, the Spotted Joe-Pye-Weed grows up to 1.5 m tall. Its leaves are in whorls of fours, and are lance-shaped, toothed and stalked. The compact, terminal, flattened flowerhead is composed of many individual light purplish-pink flowers, each with thin projecting wire-like rays. The whole flowerhead is fuzzy in appearance.

It is found on riverbanks and wet locations in the late summer and fall.

Michael Collins

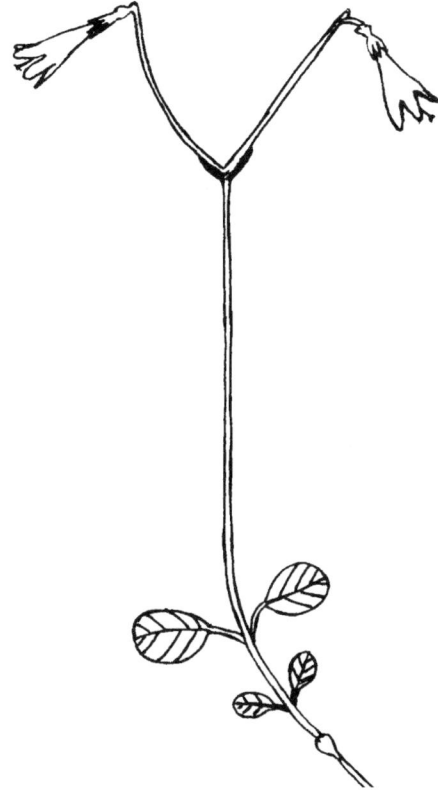

TWINFLOWER
Linnaea borealis
Honeysuckle family

The Twinflower is a delicate wildflower which is the symbol of the University's Botanical Garden at Oxen Pond. It is a semi-woody evergreen shrub which grows up to 10 cm in height. The opposite leaves are small and rounded, and it has two pink, bell-like flowers at the top of a long vertical stalk.

It is found in forested areas and shaded locations during the summer.

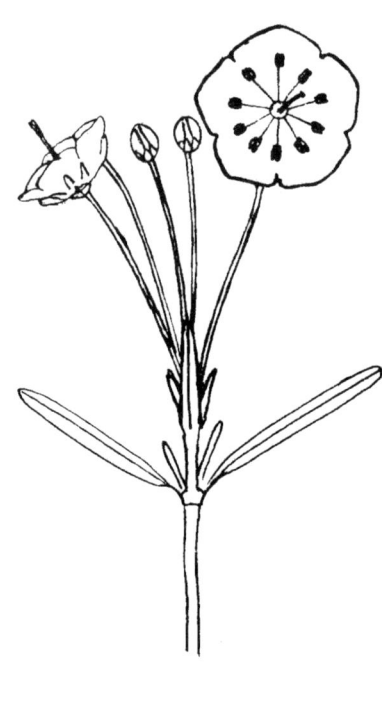

Leaf

BOG LAUREL
Kalmia polifolia
Heath family

This is a small, woody shrub of damp areas which grows up to 60 cm tall. The leaves are opposite, evergreen, long and narrow, dark green above and very light green below. Pinkish saucer-shaped flowers arise on separate, long stalks above the leaves, and each flower has five petals fused into a saucer-shaped disc.The ten stalked stamens are flattened against the petals, and the stigma projects outward from the centre of the flower which grows to 15 mm in diameter.

It is found in damp peaty areas in the early summer.

(Sheep Laurel has leaves in whorls of threes.)

Michael Collins

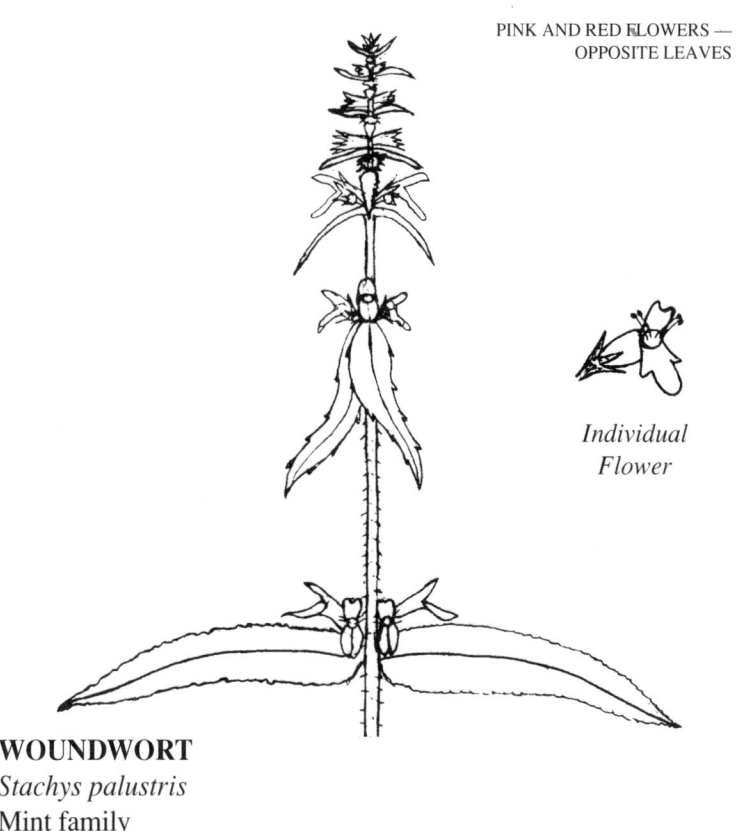

*Individual
Flower*

WOUNDWORT
Stachys palustris
Mint family

The woundwort is a tall, erect herb of damp areas which grows to 1 m in height. Its bristled stems, square in cross section, range from green to dark purple in colour. The opposite leaves are long and pointed, and roughly lance-shaped with small teeth, and are usually stalkless. The hooded pink flowers, with three-lobed lower lips, grow in circular clusters among the leaves.

It is found in meadows, ditches and other damp locations in the late summer to fall.

(Hemp Nettle has swollen stems below the leaf attachedments.)

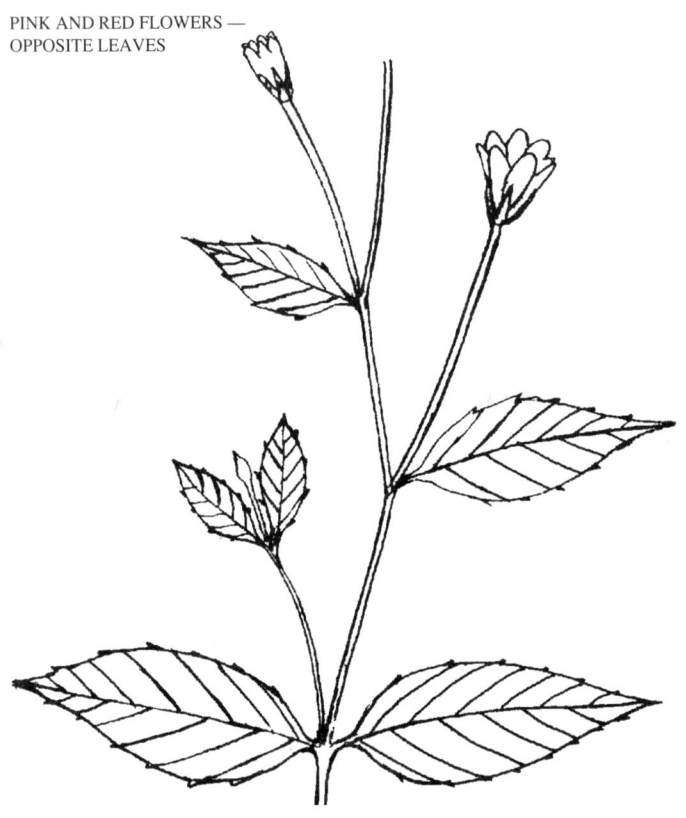

NORTHERN WILLOWHERB
Epilobium glandulosum
Evening primrose family

The Northern Willowherb is an erect herb which grows to 1 m in height. Its alternate upper leaves, with the lower opposite, are elliptical and toothed with very short stalks. The flowers are pink, four-petalled, and appear to have long, narrow stalks, which actually contain the ovaries. After fertilization long narrow seed pods develop from the ovaries. ·

It is found in shaded areas in the summer and fall.

Michael Collins

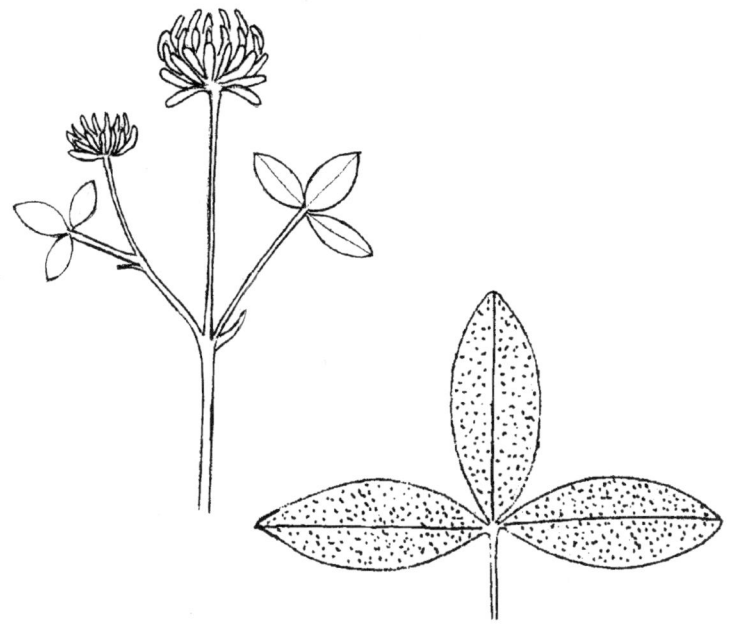

ALSIKE CLOVER
(Alsatian Clover)
Trifolium hybridum
Pea family

This clover is a low growing herb, up to 30 cm in height, of open areas with leaves and flowers arising from the same stalks. It has the familiar clover-type tripartite compound leaves, but are not marked with whitish chevrons as in the Red and White Clovers. Spherical compact flowerheads are each composed of several dozen individual flowers, white or pinkish in colour.

It is found in fields, roadsides, and waste ground during the summer and fall.

(Red Clover has leaf chevrons; White Clover has leaves and flowers arising on separate stalks.)

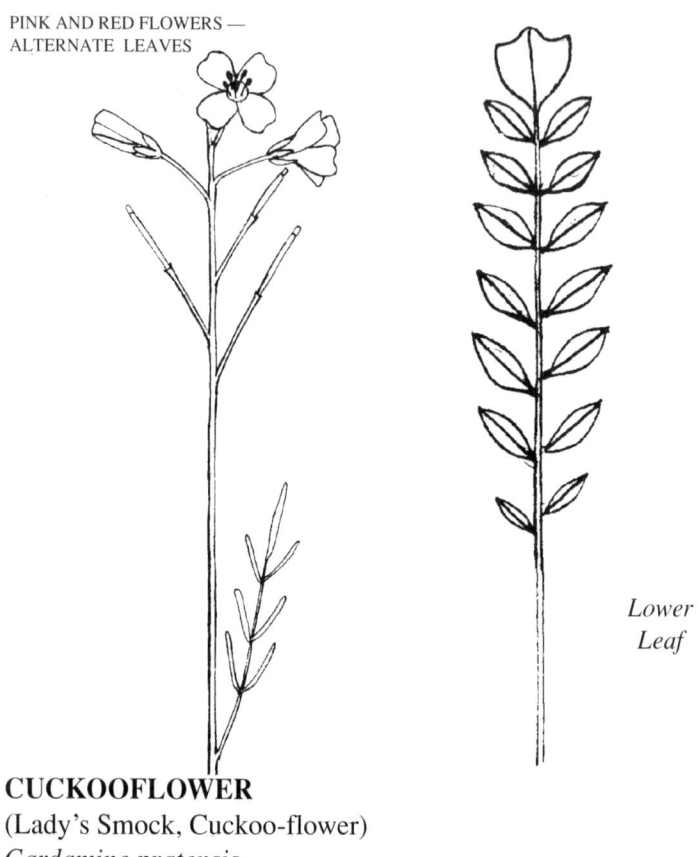

*Lower
Leaf*

CUCKOOFLOWER
(Lady's Smock, Cuckoo-flower)
Cardamine pratensis
Mustard family

An erect herb of damp areas, the Cuckcooflower grows to
1 m in height. Leaves are alternate and compound, with many
paired ovate leaflets on the lower leaves. The leaflets on the
upper leaves are narrow. Numbers of stalked, four-petalled pink,
sometimes white, flowers grow at the tip of the stem, spreading
to 15 mm in diameter.

It is found on riverbanks, in meadows, and in other damp
locations in the late spring and early summer.

Michael Collins

NORTHEASTERN ROSE
Rosa nitida
Rose family

This is a branching, bushy shrub which grows to 1 m in height. Its stems are covered with small bristles, and the leaves are compound with finely toothed leaflets. It has fragrant flowers which are five-petalled and large, spreading up to 5 cm across, and light pink in colour with sepals projecting out beyond the petals. Red fruits called 'hips' are produced in the fall.

It is found on riverbanks, ditches and other damp locations in the late summer.

(Virginian Rose has scattered prickles.)

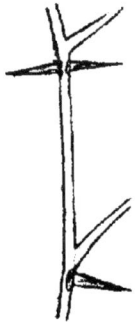

Stem, showing prickles

VIRGINIAN ROSE
Rosa virginiana
Rose family

This rose is similar to the Northeastern Rose but the stems have scattered stout prickles rather than bristles. It has distinctive, fragrant, pink rose-type flowers, compound leaves, and stems with stout prickles.

It is found in drier as well as damper locations.

Michael Collins

Leaf

PARTRIDGEBERRY
(Mountain Cranberry)
Vaccinium vitis-idaea
Heath family

The Partridgeberry is a ground-hugging evergreen shrub of exposed areas and can form extensive carpets. The leaves alternate, are minute, and grow up to 6 mm long, with dark green leaves above, and paler below. Clusters of small, pink, drooping, bell-shaped flowers grow at the tip of the stem. Edible wine-red berries are produced in the fall.

It is found on barrens and in exposed areas in the summer.

(Blueberries are taller with deciduous leaves.)

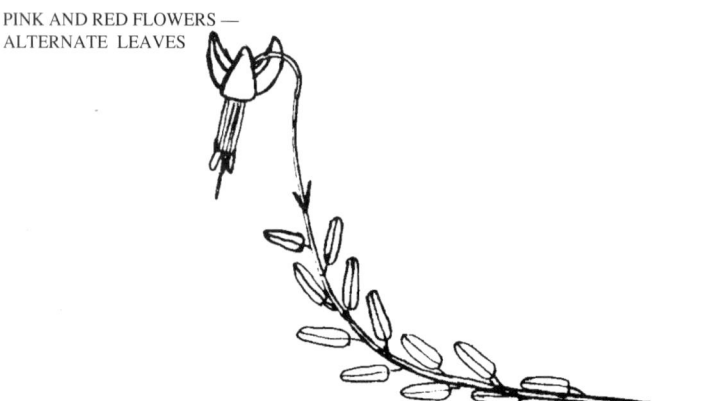

CRANBERRIES
Vaccinium species
Heath family

Cranberries are small, creeping evergreen shrubs of boggy areas, growing up to 8 cm in height. Leaves are alternate, evergreen and minute, and are green above and whitish below. It has two small, drooping flowers which grow at the tips of long erect stalks. The four petals of the flower are curved back upwards, giving a resemblance to a 'Turk's cap'. The stamens form a hollow tube surrounding the single style. Red, edible berries are produced in late fall which are best picked the following spring after overwintering.

It is found in bogs, and other damp areas.

CRANBERRY (not illustrated)
(Large Cranberry)
Vaccinium macrocarpon
Its flowers are not terminal, but arise from behind the leaves.

MARSHBERRY
(Small Cranberry)
Vaccinium oxycoccus
Its flowers are terminal, and arise from in front of the leaves.

Michael Collins

MUSK MALLOW
Malva moschata
Mallow family

See description under white flowers. Flowers can be either white or pink.

COMMON BURDOCK
Arctium minus
Daisy family

This is a tall, erect, bushy herb, which grows to 1 m in height. Its stems are thick and ridged, with long-stalked leaves; the lower ones heart-shaped. Flowerheads are stalked, thistle-like bristly burs, which spread up to 2 cm in diameter. Open flowers have lavender rays projecting outward, and white pollen-covered stamens. After fertilization the flowerheads develop into prickly, clinging burs.

It is found on roadsides and in waste places during the summer and fall.

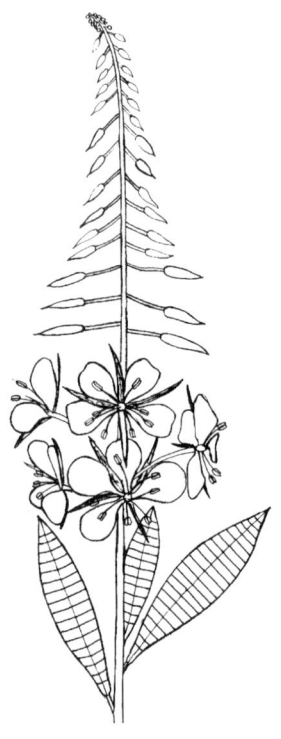

FIREWEED
(Great Willowherb)
Epilobium angustifolium
Evening-primrose family

    The Fireweed is a tall herb which grows up to 1.5 m; often in dense stands. Its stems are crowded with alternate, elliptical, stalked leaves. The flower buds are on long, drooping stalks near the stem tip. It has many pink, four-petalled flowers, spreading to 3 cm across in a spike above the leaves with the sepals projecting beyond the petals. Flowers produce long, thin seed pods which angle upward.

    It is found on roadsides, burnt-over areas, and waste ground during the late summer and fall.

Michael Collins

BOG-ROSEMARY
(Bog Rosemary)
Andromeda glaucophylla
Heath family

   The Bog Rosemary is a low growing, up to 30 cm in height, woody shrub. Its leaves are alternate, narrow and linear, green above and whitish below. It produces flowers which are small, urn-shaped, and nodding.
   It is found in bogs during early summer.

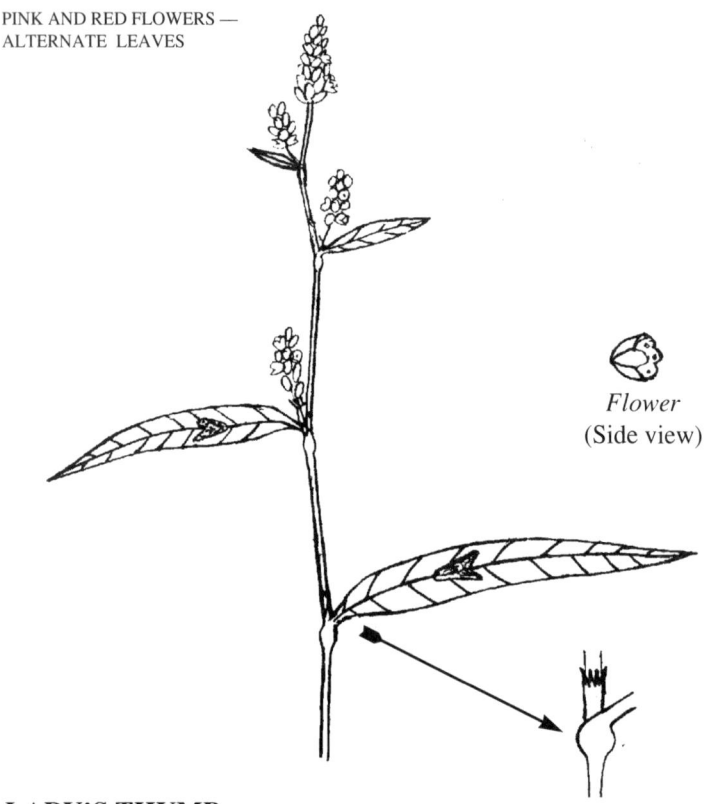

Flower
(Side view)

LADY'S THUMB
(Redleg)
Polygonum persicaria
Buckwheat family

This is an erect herb with clusters of minute flowers. Stems are swollen just below points of leaf attachment. Leaves are lance-shaped, and often with dark blotches. Papery sheaths with fringes surround the stem just above the points of leaf attachment. It has several clusters of small white or pink flowers, each spreading to 3 mm across.

It is found in gardens and waste places during the summer and fall.

Michael Collins

Lilac, Mauve, Purple, and Violet Flowers

No Leaves at Flowering Time

RHODORA
Rhodora canadense
Heath family

This is an erect, woody shrub which grows to 1 m tall. Flowers appear before the leaves expand. The leaves are elliptical, and arise beneath the flowers. It has a terminal cluster of prominent, large (4 cm), two-lipped flowers; upper lip is two or three lobed, often deeply cleft; lower lobe is two-lobed and very deeply cleft. Lobes may be so deeply cleft as to give the appearance of having four or five separate petals. There are ten stamens and one stigma, each on a long stalk projecting outward from the flower.

It is found in bogs, heathlands, pond edges, and edges of woods in the early summer.

Leaves in Basal Rosette

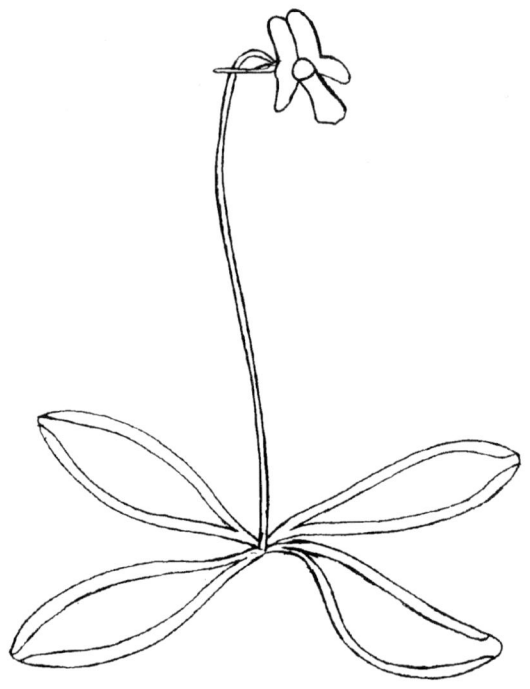

BUTTERWORT
Pinguicula vulgaris
Bladderwort family

A low growing insectivorous herb growing up to 25 cm tall, it has a basal rosette of yellowy-green, greasy leaves with inrolled edges. Insects stick to the greasy surfaces and then become digested. The greasy nature of the leaves gives rise to the common name. It has a single five-lobed violet flower with a backward-pointing spur at the tip of a long stalk.

It is found in damp areas, including roadside ditches in the early summer.

Michael Collins

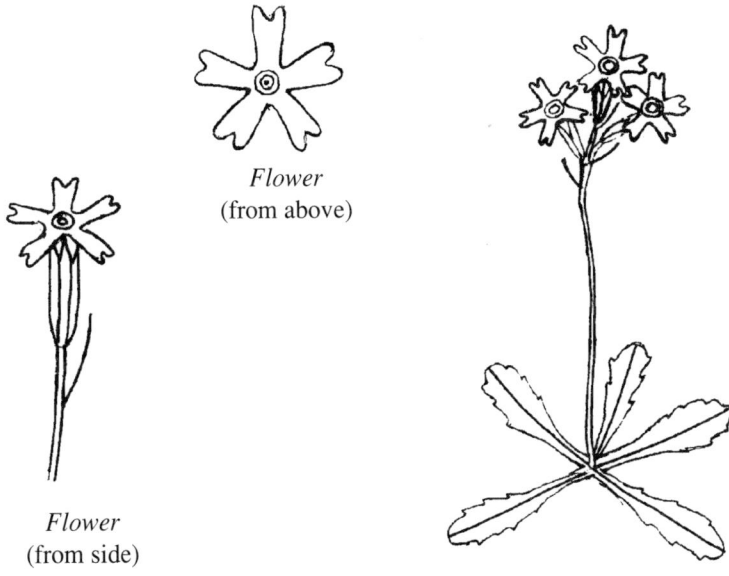

Flower
(from above)

Flower
(from side)

ST. LAWRENCE BIRD'S-EYE PRIMROSE
(Bird's-eye Primrose)
Primula laurentiana
Primrose family

A small, erect, low growing herb which grows to 10 cm tall, it has a basal rosette of slightly toothed, elliptical leaves. Its terminal cluster of five-petalled, violet flowers, spreading to 1 cm across, have a central yellow ring. Petals are slightly indented at the tips, and each flower has a single bract at its base.

It is found in open, dry calcareous areas (common along the Northern Peninsula highway), in the early summer.

Plants and Wildflowers of Newfoundland   121

Leaves Alternate — Compound

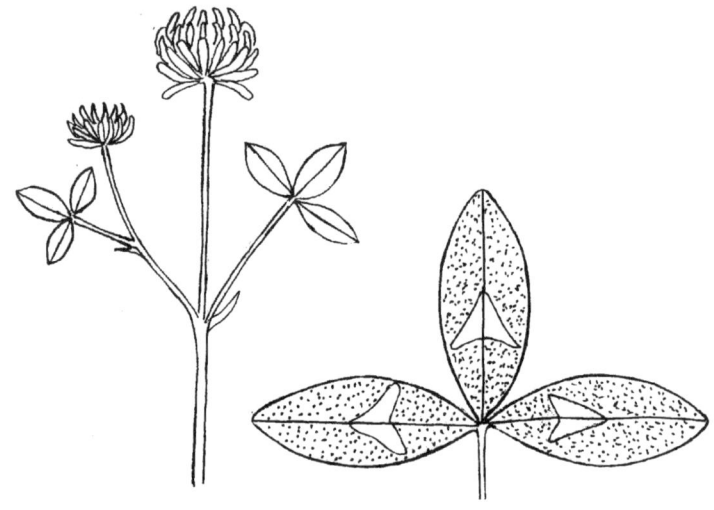

RED CLOVER
Trifolium pratense
Pea family

Red Clover Leaf
(showing pale chevrons)

This is a low growing herb of open areas. The leaves and flowers arise from the same stalks. It has the familiar clover-type, compound leaves with three leaflets, rarely four, with each leaflet marked with a whitish chevron. Spherical compact flowerheads, each composed of several dozen individual flowers, are purple in colour. Surely one of our most inappropriately named wildflowers since the flowers are purple and not red as the name implies!

It is found on lawns, fields, roadsides and waste places in the summer and fall.

(Alsike clover does not show chevrons.)

Michael Collins

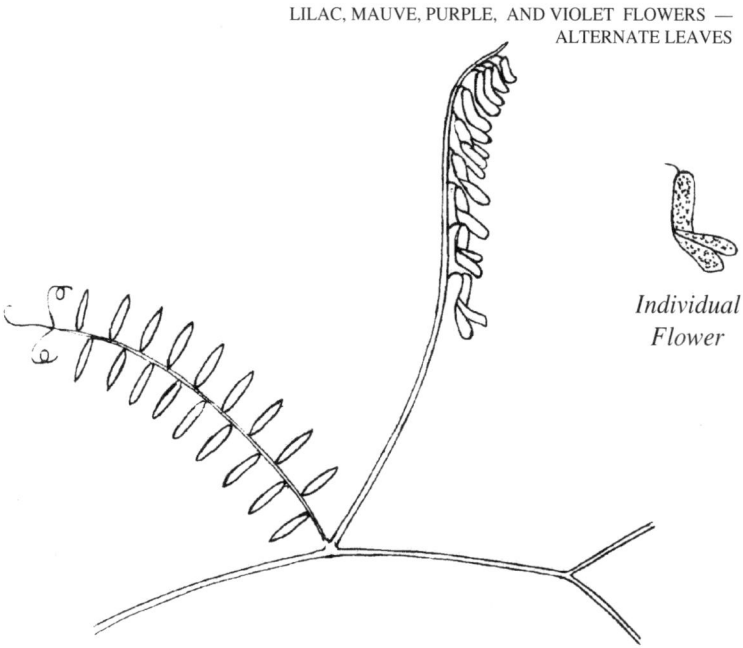

*Individual
Flower*

COW VETCH
Vicia cracca
Pea family

A straggling herb, the Cow Vetch uses leaf tendrils to fasten on to other plants, posts and the like to support the plant. The leaves are compound with three tendrils at the tip which curl around plants and other objects for support. Its flowers, dark purple in colour, grow on separate stalks from the leaves. Individual flowers usually grow in a line along one side of the flower stalk.

It is found in grassy areas during summer and fall.

Leaves Alternate — Simple

CANADA THISTLE
(Creeping Thistle)
Cirsium arvense
Daisy family

The Canada Thistle is a tall, erect herb which grows up to 1 m in height. Its leaves are stiff and crinkled with many sharp spines. Many lilac flowers are produced which spread to 2 cm across.

It is found in fields, roadsides and waste places during the late summer and fall.

(Bull Thistle has spines on both leaves and stem.)

Michael Collins

BULL THISTLE
(Stinger Nettle)
Cirsium vulgare
Daisy family

This is a tall, erect herb, growing up to 1.5 m in height. The leaves are stiff and very spiny, and leaf spines continue along ridges on the stem. Several compact, large violet flowers are produced which grow up to 4 cm in height.

It is found in fields, roadsides and waste places in the late summer and fall.

(Canada Thistle has spines only on the leaves.)

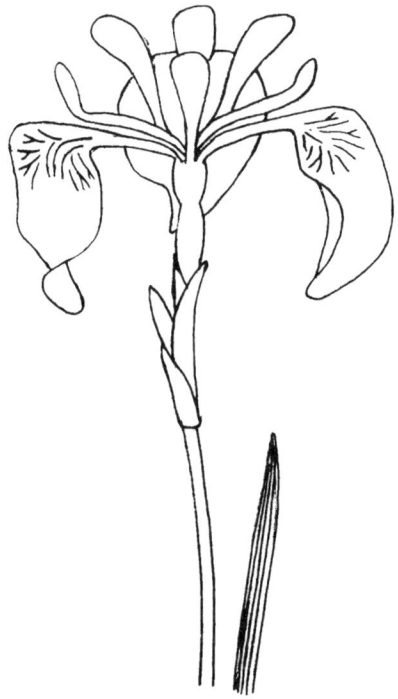

BLUE FLAG
(Wild Iris, Larger Blue Flag)
Iris versicolor
Iris family

The Blue Flag is a tall herb of wet areas, which grows up to 1 m in height. Long, linear leaves sheath the stem and it bears several, large, three-petalled, showy, violet flowers. The larger sepals curve down and have prominent yellow markings. Flowers spread to 10 cm in diameter.

It is found in bogs and other wet places in the summer.

(A similar species, Yellow Iris (*Iris pseudacorus*) has yellow flowers.)

Michael Collins

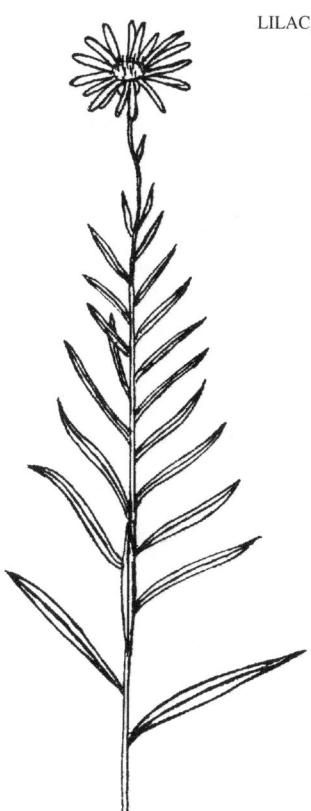

BOG ASTER
Aster nemoralis
Daisy family

An erect herb of damp areas which grows up to 50 cm in height, the Bog Aster has numerous grass-like leaves which crowd the thin slender stem. Usually one solitary light violet, rayed flower grows at the top of the stem which spreads to 3 cm across.

It is found in bogs and damp areas in the late summer and early fall.

(Other asters have more than one flowerhead.)

Plants and Wildflowers of Newfoundland   127

HAREBELL
Campanula rotundifolia
Bluebell family

An erect herb of exposed areas which grows up to 20 cm in height, the Harebell has grass-like leaves. It bears several violet bell-shaped flowers at the top of the stem which spread to 2 cm across.

It is found in exposed grassy areas, such as clifftops, in the late summer.

128   Michael Collins

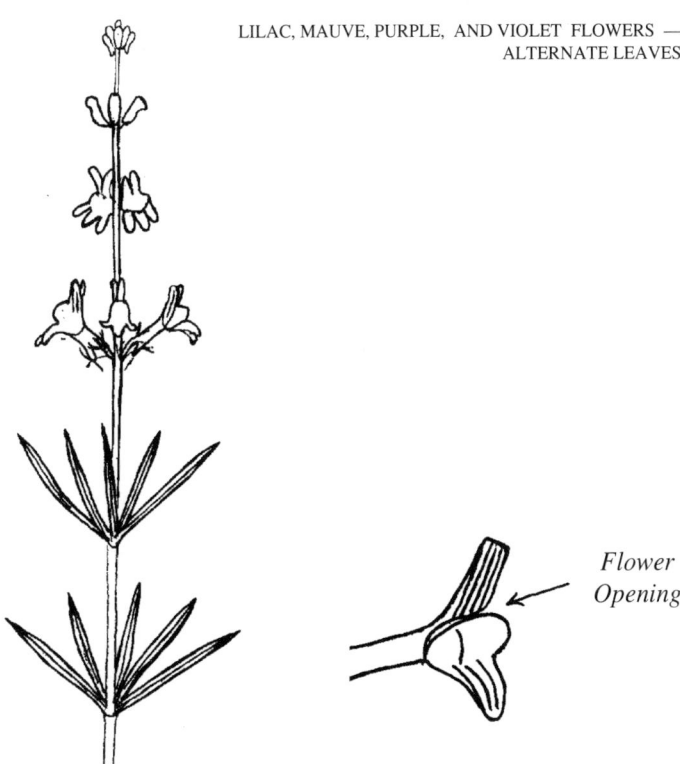

*Flower
Opening*

STRIPED TOADFLAX
Linaria repens
Snapdragon family

This is an erect herb of open areas, growing up to 50 cm tall. The stem is crowded with alternate, linear leaves which are sometimes so closely grouped as to appear whorled. Light violet, irregular, snapdragon-like flowers grow in circular clusters above the leaves. Petals of the flowers show darker purple guidelines and each flower has a short spur.

It is found on roadsides, and waste places from summer to fall.

(Butter-and-eggs has yellow and orange flowers.)

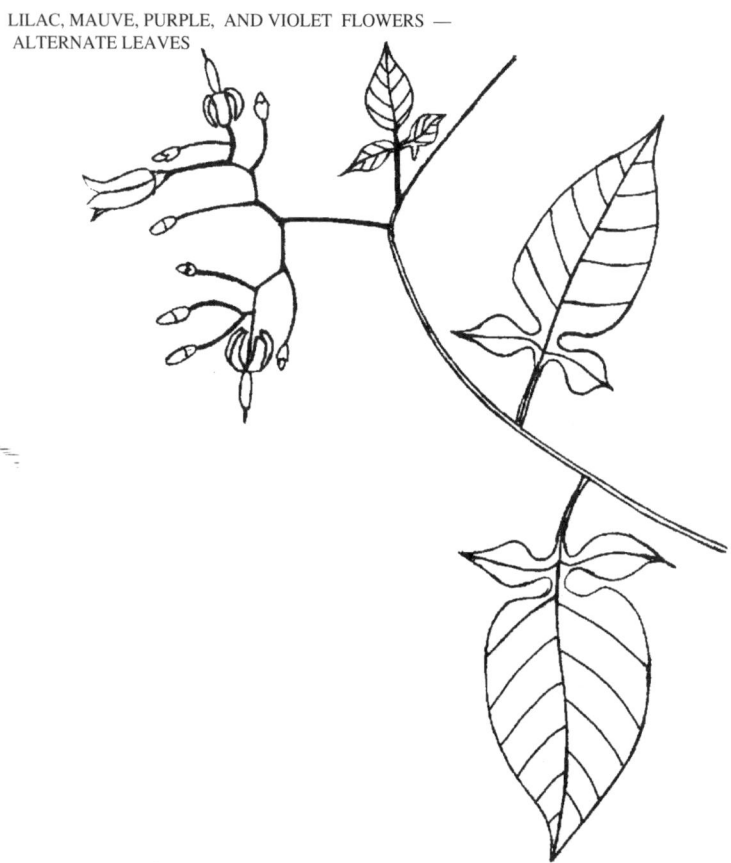

NIGHTSHADE
(Bittersweet)
Solanum dulcamara
Tomato family

A weak-stemmed climbing vine, the Nightshade has heart-shaped leaves with two distinctive small basal lobes. It bears a large number of long-stalked flowers, each with five swept back violet petals, and a protruding beak of yellow anthers. It also produces hanging clusters of deep red berries.

It is found in moist areas in the late summer.

Michael Collins

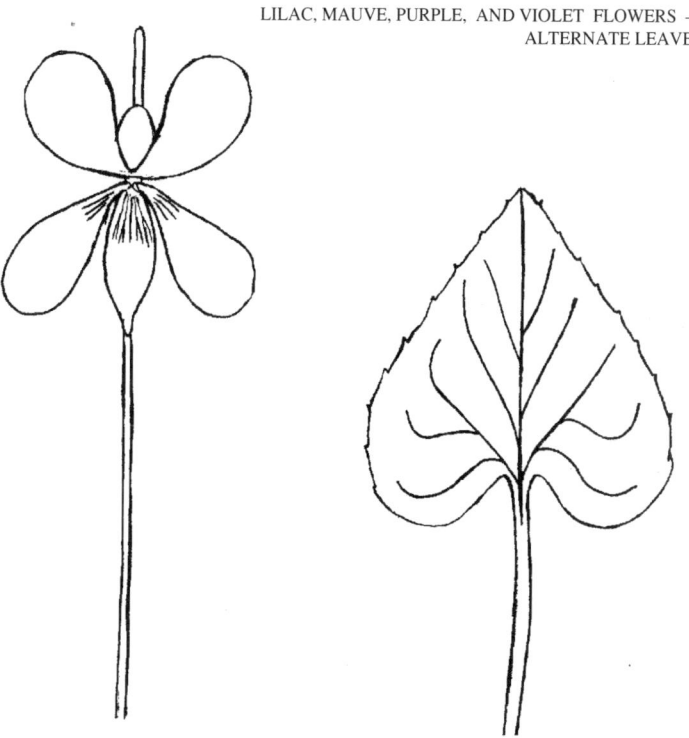

MARSH BLUE VIOLET
Viola cucullata
Violet family

The Marsh Blue Violet is an herb of damp areas. Flowers and leaves grow on separate, long stalks up to 20 cm tall. Its leaves are heart-shaped and long-stalked. Flowers are violet in colour and have five petals, two upper and three lower, the latter with noticeable veins, and spread up to 2 cm in diameter.

It is found in meadows, bogs, streambanks, and other damp locations in the spring.

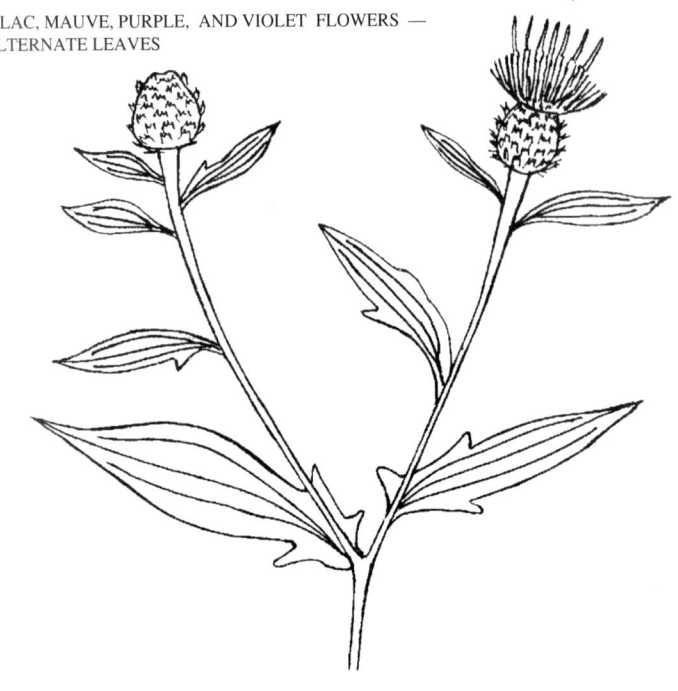

BLACK KNAPWEED   (Hardheads)
Centaurea nigra
Daisy family

An erect herb, the Black Knapweed grows up to 1 m in
height. Its leaves are lanceolate with prominent toothlike projections near bases of lower leaves. The compact flowerhead is thistle-like, and usually violet in colour, but sometimes light pink,
spreading 25 mm across. Its stalk is swollen just below the flowerhead. The base of the flowerhead is covered with brown, scale-
like projections. Old flowerheads (hardheads) persist through
winter, often harbouring overwintering insect larvae. These
infected heads are harder than the uninfected.

It is found in fields, lawns, and roadsides in the late summer
and fall.

(Thistles possess spores.)

Michael Collins

LILAC, MAUVE, PURPLE, AND
VIOLET FLOWERS —
ALTERNATE LEAVES

PURPLE-STEMMED ASTER
*Aster puniceu*s
Daisy family

A very tall herb of wet areas, the Purple-stemmed Aster grows up to 2 m in height. The stem is usually purplish and bristly. Its leaves are large, pointed and lance-shaped, and are crowded on the stem. The leaf base clearly clasps the stem. It bears large, many rayed (about 40) violet flowerheads which spread to 3 cm in diameter.

It is found on riverbanks, ditches and other wet areas in the late summer and fall.

(Other asters are not as tall and their leaves are not as clasping.)

NEW YORK ASTER
Aster novae-belgii
Daisy family

This is an herb of damp areas, usually growing to less than 1 m in height. Its stem is crowded with leaves, which are not purplish or bristly. The leaves, usually very long and narrow, sometimes slightly toothed, only slightly clasp the stem. Many violet flowerheads are produced, usually with 30 or more rays, and the floral bracts reflexed.

It is found in ditches and other damp areas during late summer and fall.

(Purple-stemmed Aster has bristly stems, and noticeable clasping leaves; Rough-leaved Aster has fewer flower rays and shorter leaves.)

Michael Collins

ROUGH-LEAVED ASTER
Aster radula
Daisy family

The Rough-leaved Aster is an erect herb, which grows up to 1 m in height. Its leaves are long, pointed lanceolate, sometimes toothed, only slightly clasping the stem, if at all. Several, many-rayed (about 20) violet flowerheads with greenish-yellow discs are produced.

It is found in meadows and moist shady areas, usually in late summer and fall.

(The Purple-stemmed and New York Asters have many more rays on each flowerhead and larger leaves.)

BLUE FLOWERS

Leaves in A Basal Rosette

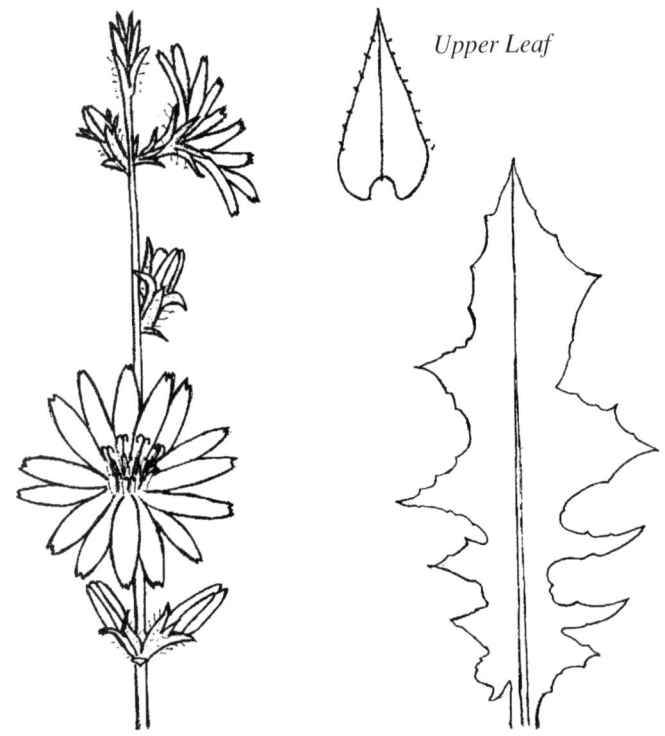

Upper Leaf

CHICORY
Cichorium intybus
Daisy family

This is an erect herb which grows up to 1.5 m tall. Its basal leaves are dandelion-like in a basal rosette. It bears stalkless, blue dandelion-type flowerheads.

It is found on roadsides and waste places in the summer and fall.

Michael Collins

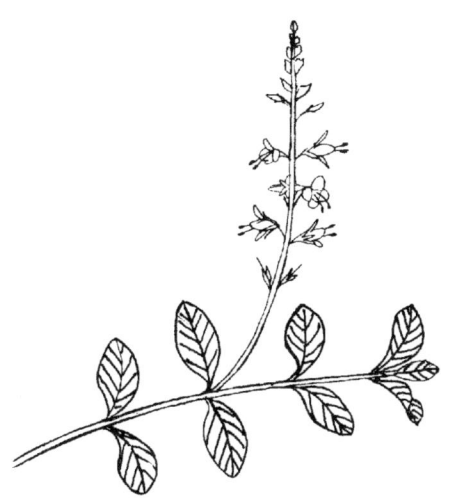

Flower
(from above)

COMMON SPEEDWELL
Veronica officinalis
Snapdragon family

A small, low growing herb, up to 20 cm in height, the Common Speedwell has oval opposite leaves. Pairs of small, blue flowers arise from an erect stalk. Each flower has four petals, each with darker stripes, and two stamens; the lowest of the four petals is the smallest. These flowers spread to 7 mm in diameter.

It is found in fields and waste places from the spring to fall.

Leaves Alternate

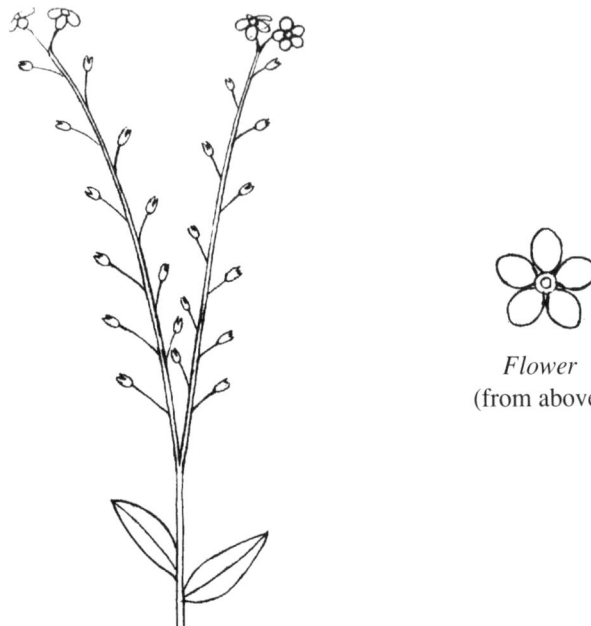

Flower
(from above)

TRUE FORGET-ME-NOT
Myosotis scorpioides
Forget-me-not family

The True Forget-me-not is an erect, low growing herb, which grows up to 50 cm tall. The alternate leaves are sessile and elliptical. Many small flowers arise from two erect, diverging stalks, each flower having five blue petals and a distinctive yellow 'eye'. True Forget-me-not has larger flowers (to 8 mm across) than other species, lives in damper areas, and bears no leaves on the flower stalks. There are a number of other species present in the province, but each has the same type of flower as the True Forget-me-not with the central yellow 'eye'.

They are found in gardens, streambanks, and other wet places in the spring to fall.

138   Michael Collins

ORANGE FLOWERS

Leaves in A Basal Rosette

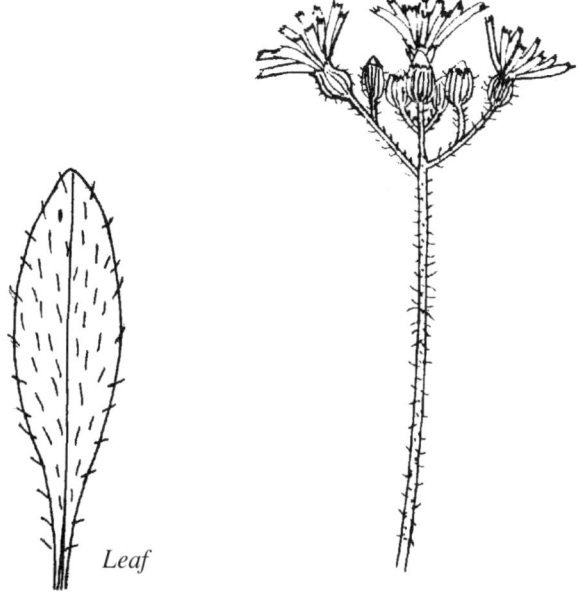

Leaf

ORANGE HAWKWEED
(Devil's Paintbrush)
Hieracium aurantiacum
Daisy family

This is an erect dandelion-like plant with orange flowers, and is our only native plant with fully orange flowers. It grows up to 50 cm tall. The stalk and flower bases are covered with black hairs with a basal rosette of elliptical leaves. Several bright orange dandelion-like flowers are borne at the tip of the stalk.

It is found on roadsides and waste places in the summer and fall.

Leaves Alternate — Compound

BIRDFOOT TREFOIL
Lotus corniculatus

(See description under yellow flowers.)

Leaves Alternate — Simple

BUTTER-AND-EGGS
Linaria vulgaris

(See description under yellow flowers.)

Michael Collins

GLOSSARY OF TERMS

alternate (branches, buds, flowers, leaves) not opposite one another; buds, leaves, branches and flowers arise singly at different points on the stem

annual surviving only one season

anther the enlarged upper part of the stamen containing the pollen grains

aromatic producing an aroma or scent

ascending growing upward

axil the angle between the stem and the leaf stalk

basal situated at the bottom of the stem

berry a fleshy fruit usually containing a number of small seeds

blade the flat, expanded part of the leaf

bract a small leaf, green or other colour, usually located at the base of the flower

bristle a stiff, pointed, hair-like projection

bur a bristle-covered fruit which often becomes attached to passing animals

calyx the outermost circle of flower parts, made up of sepals; usually green but sometimes coloured and petal-like

clasping (leaf) partly surrounding the stem

cleft (leaf, petal) deeply cut

compound (leaf) divided into smaller, separate, leaflets

corolla the circle of flower parts inside of the calyx; made up of petals which are usually coloured

creeping (stem) running along over the ground, and rooting at intervals

deciduous (leaves) falling off at the end of the season; not persistent

disc (flower) the central part of the head in composite flowers (members of the Daisy family); made up of many, small, tubular disk flowers; usually surrounded by a circle of flat ray flowers

dissected (leaf) divided into many narrow segments, but not compound

divided (leaf) cleft to, or almost to, the base or midrib
                        divided into many narrow segments, but not compound

downy covered with soft, very fine hairs

elliptical (leaf) broadest in the middle, and tapering toward both the top and bottom ends

entire (leaf) the leaf margin is continuous, without lobes, teeth and the like

erect upright or vertical

evergreen the leaves remain greenish in colour, and stay on the plant during the winter

family a group of related plants

filament the stalk of the stamen which bears the anthers at its tip

genus a group of closely related species. The genus (or generic name) is the first of the two biological (Latin) names. The first letter of the generic name is always capitalized

gland a minute swelling, like a swollen hair, which secretes oils or other substances

guide line a darker stripe on a petal or sepal which serves as a direction marker for insect pollinators

head (or flowerhead) a group of flowers joined into a crowded terminal cluster; as found in clovers and composites (members of the Daisy family)

herb a non-woody plant

hood in a number of flowers, particularly members of the Mint family, the upper part of the corolla resembles a hood

inflated swollen and balloon-like

insectivorous insect-eating

indented (leaf) the leaf margin is slightly notched inward

inrolled (leaf) the edges of the leaf are rolled inward

introduced not native to the area

irregular (flower) not symmetrical; the parts of the flower are dissimilar in shape and size

lance-shaped (leaf) broader toward one end and tapering to the other end; usually three or more times longer than broad

leaflet one of the separate leaf-like parts of a compound leaf

linear (leaf) long and narrow; the leaf veins are parallel

lip the upper or lower part of the corolla (petals) in a number of irregular flowers

lobe one of the segments, usually rounded, of a leaf or a flower

oblong (leaf) longer than broad, with parallel sides

opposite (branch, bud, flower, leaf) arranged in pairs on opposite sides of the stem

ovate almost oval

ovary the enlarged base of the pistil which contains the immature seeds

palmate (leaf) hand-like, with the leaflets radiating outward from a central point

perennial a plant usually living more than two years

petal one of the individual parts of the corolla; usually coloured

petiole the leaf stalk

pistil the central female reproductive part of the flower; composed of the ovary, style, and stigma

pollen the male spores; produced by the anther

prickle a small, wood-like stem projection, tapering from a broad base to the sharp tip

prostrate laying flat on the ground

ray (ray flower) the flat strap-like flowers that surround the central disc in composite flowers (members of the Daisy family)

reflexed abruptly turned backward or downward

regular (flower) all the different parts of the flower are identical in shape and in size to each other

rosette (leaf) a circle of leaves arising so closely to one another as to appear to be in a whorl; usually at the base of the stem

runner a slender, prostrate branch which can send out roots and leaves to produce a new plant

scale a tiny colourless leaf found on some stems

sepal one of the individual parts of the calyx; usually green in colour

sessile (flower, leaf) not having a stalk

sheath a thin membrane surrounding the stem

shrub a woody plant usually less than 3 m in height, and often with several main stems

simple (leaf) not compound; not divided into leaflets

species a distinct kind of plant; members of the same species can interbreed, but members of different species rarely can. The species (or specific) name is the second of the biological (Latin) names, coming after the generic name.

spike an elongated, often vertical, cluster of flowers in which the stalkless or short-stalked flowers arise from a central stem

spur a hollow, tubular projection of a flower; often contains nectar

stalk (flower, leaf) the stem of a flower or leaf

stamen the male organ of the flower consisting of the filament and the pollen-containing anther; usually several in number

stigma the tip of the pistil which receives the pollen grains from another flower

stipule a small, leaf-like growth at the base of the petiole in some plants; usually found in pairs

style the stalk of the pistil which bears the stigma at its tip

tendril a slender, coiling extension of a leaf used for climbing and support

terminal at the end of a stem or branch

toothed (leaf) having tooth-like projections along the leaf margin

trailing running along over the ground but not rooting

translucent allowing light to pass through it; almost, but not, transparent

tree a woody plant usually more than 3 m in height, and usually having one main stem

tuber a short, thick underground stem

vein a tube running through a leaf; conducts water and nutrients to and from the stem

vine a woody, climbing stem twining around other plants and objects for support

whorl three or more leaves arranged in a circle around the stem

wing a thin, narrow membrane extending along a stem or other part of a plant

REFERENCES

"Terra Nova National Park wildflowers checklist." Parks Canada, Ministry of the Environment, Ottawa, Ontario, 1983.

Barrett, W., MacKay, A. and Griffin, D. "Atlantic wildflowers". Oxford University Press, Toronto, Ontario, 1984.

Frankton, C. and Mulligan, G.A. "Weeds of Canada". Publication 948. Canada Department of Agriculture, Ottawa, Ontario, 1974.

Mann, H. "Scientific and common names of Newfoundland vascular plants listed alphabetically by genus". Biology Department, Sir Wilfred Grenfell College, Corner Brook, Newfoundland, 1991.

Meades, W.J. and Moores, L. "Forest site classification manual. A field guide to the Dammar forest types of Newfoundland." FRDA Report 003. Forestry Canada, Newfoundland and Labrador Region, St. John's, Newfoundland, 1989.

Newcomb, L. "Newcomb's wildflower guide". Little, Brown and Company, Boston, Massachusetts, 1977.

Peterson, R.T. and McKenny, M. "A field guide to wildflowers of northeastern and north-central North America." (The Peterson Field Guide Series). Houghton-Mifflin Company, Boston, Massachusetts, 1968.

Rouleau, E. "A checklist of the vascular plants of the province of Newfoundland". Contribution de l'Institut Botanique de l'Universite de Montreal, Montreal, Quebec, 1956.

Ryan, A.G. "Common flowers of the forest floor". Park Interpretation Publication Number 10, Parks Division, Department of Tourism,

Government of Newfoundland and Labrador, St. John's, Newfoundland, 1973.

Ryan, A.G. "Native trees and shrubs of Newfoundland and Labrador". Parks Division, Department of Culture, Recreation and Youth, Government of Newfoundland and Labrador, St. John's, Newfoundland, 1978.

Ryan, A.G. "Some peatland flowers." Parks Interpretation Pamphlet #13. Parks Division, Department of Tourism, Government of Newfoundland and Labrador, St. John's, Newfoundland, 1978.

Scott, P.J. "Flosculous Snippets". (Biological and common names of Newfoundland plants). Osprey, 8, 1, 1-29, 1977.

Scott, P.J. "Biology 3041 Boreal Flora—Vascular Flora of Newfoundland (Laboratory Manual)". Memorial University of Newfoundland, St. John's, Newfoundland, 1993.

Semple, J.C. and Heard, S.B. "The Asters of Ontario: Aster and Virgulus Raf. (Compositae: Astereae)". University of Waterloo Biology Series No. 30. Department of Biology, University of Waterloo, Waterloo, Ontario, 1987.

Semple, J.C. and Ringius, G.S. "Goldenrods of Ontario: Solidago and Euthamia Nutt." Revised Edition by John C. Semple. University of Waterloo Biology Series No. 36. Department of Biology, University of Waterloo, Waterloo, Ontario, 1992.

Venning, F.D. "A guide to field identification—wildflowers of North America". Golden Press, New York, New York, 1984.

Watts, M. "Flower finder: a guide to identification of spring wild flowers and flower families". Nature Study Guild, Berkeley, California, 1954.

Zim, H.S. and Martin, A.C. "Flowers—a guide to familiar American wildflowers". Golden Press, New York, New York, 1950.

INDEX

Michael Collins